The Conquest of Time

and

The Happy Turning

BY THE SAME AUTHOR
ALL PUBLISHED BY HOUSE OF STRATUS

FICTION

ANN VERONICA
APROPOS OF DOLORES
THE AUTOCRACY OF MR PARHAM
BABES IN THE DARKLING WOOD
BEALBY
THE BROTHERS *AND*
    THE CROQUET PLAYER
BRYNHILD
THE BULPINGTON OF BLUP
THE DREAM
THE FIRST MEN IN THE MOON
THE FOOD OF THE GODS
THE HISTORY OF MR POLLY
THE HOLY TERROR
IN THE DAYS OF THE COMET
THE INVISIBLE MAN
THE ISLAND OF DR MOREAU
KIPPS: THE STORY OF A SIMPLE
    SOUL
LOVE AND MR LEWISHAM
MARRIAGE
MEANWHILE
MEN LIKE GODS
A MODERN UTOPIA
MR BRITLING SEES IT THROUGH
THE NEW MACHIAVELLI
THE PASSIONATE FRIENDS
THE SEA LADY
THE SHAPE OF THINGS TO COME
THE TIME MACHINE

TONO-BUNGAY
THE UNDYING FIRE
THE WAR IN THE AIR
THE WAR OF THE WORLDS
THE WHEELS OF CHANCE
WHEN THE SLEEPER WAKES
THE WIFE OF SIR ISAAC HARMAN
THE WONDERFUL VISIT
THE WORLD OF WILLIAM CLISSOLD
    VOLUMES 1,2,3

NON-FICTION

EXPERIMENT IN AUTOBIOGRAPHY
    VOLUMES 1,2
H G WELLS IN LOVE
THE OPEN CONSPIRACY AND OTHER
    WRITINGS

The Conquest of Time

and

The Happy Turning

H G WELLS

HOUSE OF
STRATUS

The Conquest of Time First Published 1942
The Happy Turning First Published 1945

Copyright by The Executors of the Estate of H G Wells

This edition published in 2002 by House of Stratus, an imprint of House of Stratus Ltd, Thirsk Industrial Park, York Road, Thirsk, North Yorkshire, YO7 3BX, UK.
Also at: House of Stratus Inc., 2 Neptune Road, Poughkeepsie, NY 12601, USA.

www.houseofstratus.com

Typeset, printed and bound by House of Stratus.

A catalogue record for this book is available from the British Library and The Library of Congress.

ISBN 0-7551-0397-1

THE CONQUEST OF TIME

CONTENTS

THE HAPPY TURNING
CONTENTS

The Conquest of Time

Prefatory Note and Dedication

Before sending this book to the press I submitted the original draft to a number of friends for whose critical intelligence I had a profound respect and who were interested in the questions it raised. They responded generously and fully. One grave fallacy and several serious inconsistencies were exposed, and they discovered so many minor instances of obscurity, overstatement, and understatement that very little of the original draft survived. I find it difficult to express my gratitude to the sixteen generous collaborators who have given their time and knowledge to this mental cleaning-up. To thank them by name would be to implicate them in the final formulae and statements, for which they are in no way responsible, because the decision to accept, modify, or reject, was necessarily mine; but I think that taken all together we have made a fairly lucid and consistent summary of modern ideas concerning the fundamentals and ultimates of existence. So very gratefully I dedicate my report to the Sixteen Anonymous Thinkers and hand it over to the Rationalist Press Association to replace the superannuated volume they have issued for so many years.

H G Wells

CHAPTER ONE

Time Disgorges

I n 1908, when I was still mentally adolescent, I wrote a book
called *First and Last Things*. It was an attempt to get my
ideas about the world and myself into some sort of order.
It was published, criticized, revised, and revised, and now for a
number of reasons I propose to reprint it no more, but to replace
it by the present volume. Partly it had overrun its date. There
was an account of modern war in it, a forecast of its methods and
social consequences, so completely in accordance with current
experience that it might have been written yesterday. No one
would believe it then, and now everybody would say "We all
know that." So why reprint it any more?

All through, the book abounded in ideas that were,
indistinctly under-stated or over-stated; good ideas I find them
for the most part, but ideas that can be approached, I have
realized, in a different, harder, and more effective fashion. And I
argued about a number of things where argument now seems to
me to be superfluous. My world was still innocent of psycho-
analysis, and I had never heard of Pavlov. Under the influence of
William James (*The Will to Believe*), I exaggerated the wilful
element in belief. Occasionally the book threw out sterile flowers,

bright ideas that led nowhere. Such excrescences I have pruned. This present book is a compacter and austerer book than its progenitor.

For reasons that will become clearer as the argument unfolds I call this book *The Conquest of Time*, and I begin with some rather paradoxical fencing about our everyday assumptions concerning time and the flight and direction of time. As the argument of the book develops, another thread of thought interweaves with it, a thread which came into my own speculations when I was criticizing J W Dunne's experiments in dreaming. This thread is the inquiry, "What precisely do we mean by Now?"

In those dream-studies of his, Dunne and a number of other people he induced to work with him would have a writing-pad by the bedside, and the investigator accustomed himself to write down, immediately upon awakening, whatever dream-stuff he found in his mind. He snatched back his dream before it faded or was rationalized. The results achieved seemed to show that dream-stuff is made up not only from recent events, bodily states, and so forth, distorted and rearranged, but also out of novel factors which *foreshadow* something which only arrives in the waking consciousness some hours or days later and which should be quite unforeseeable at the time. That is to say, there is a certain experimental case for inquiring whether the dreaming "Now" is of much greater extent, forward as well as backward, than the "Now" of the ordinary working life.

I was at my familiar writing-desk the other day, and before me were the sheets upon which I had been planning a series of little books, of which *The Conquest of Distance*, *The Conquest of Power*, *The Conquest of Hunger*, were to be the first three titles.

This is a time of depressing stresses, and these little books were projected to combat this depression, and so it was tempting to use the rhetorical word "Conquest." They were to be heartening records of human achievement; they were to help faltering spirits to the realization that, in spite of all our trouble and overstrain, a new life beyond all precedent may be opening for mankind. The balance of probability tilts in favour of such a new life. On the wall was a little calendar upon which I had marked the days on which these several instalments of work were to be begun, and the tortoise-shell perpetual calendar on my table said Sunday, October 12. I realized that there was something elusively absurd about these so positive dates. In what resided that faint flavour of absurdity?

It dawned upon me that the restriction of our thoughts within a narrow range of time was relaxing, and that dates are losing a definiteness they once possessed. The matter presented itself first as a clock paradox of a rather elementary sort.

"Let me get this clock idea clear," I said. "I am standing on the North Pole of the earth looking fixedly at the sun. It is twelve o'clock on Sunday. As I rotate with the earth, it continues to be twelve o'clock on Sunday until I have made a complete revolution. Then it becomes twelve o'clock on Monday, and so on *da capo*. But now suppose I begin to turn myself round faster than the earth is turning under my feet. Then I get to the starting-point sooner than the earth. If I have gone at the pace of an hour a day faster, then I get to the starting-point when it is only eleven o'clock on Sunday and I go on to –? Monday morning or Sunday evening? I consult a chronometer. The chronometer, I am assured, measures *real* time. What it does measure is an objective time that the confirmation by other human beings convinces me must be real and independent of myself. Monday morning, the chronometer tells me. Good. The day is shorter by an hour; that is all. I have to cut out that spare hour, just as a ship sailing eastward across the ocean puts its clocks on at midnight.

"But now suppose I accelerate my eastward movement very much more than that. The faster I go, the faster the sun returns to the zenith. Presently it is flickering past my eyes. By going fast enough I can get the day down to a chronometer hour, to half an hour, to five minutes, to a minute. Days and weeks spin past me."

"But you can't do that!" says someone with no ideas beyond the chronometer, and tries to end matters with that.

To which I retort cheerfully, "But I can; or in effect I can put you and myself in the same quandary."

A bright-minded critic asks me to modify this a little. He points out that if I am standing at the North Pole, and rotating so as to face the sun, the time does *not* continue to be twelve o'clock. He argues that if I am standing at the North Pole looking fixedly at the sun, everywhere immediately in front of me, between me and the South Pole, is at noon on Sunday. Immediately behind me it is midnight. If I feel hungry I can take a diagonal step forward and it is five o'clock and time for tea. If I feel sleepy I can step backwards and it is bedtime. At the Pole it is either no time at all or all times at once. One need not live outside this planet to discover that our habitual attitude to time has something queer about it. At the Pole it has become invalid and absurd. Perhaps my face is, as I face the sun, at high noon on Sunday, but my back is at midnight, conveniently divided to be apparently one shoulder-blade Sunday–Monday and the other

Saturday–Monday. Instead of merely turning to shorten my days, I should have to walk round the Pole in small circles.

I admit I may lack the constitution and necessary facilities for waltzing round the Pole in this fashion, but what if we have a nicely warmed and provided roundabout there? Or suppose we have a suitably constructed aeroplane and fly round the world, not precisely at the Pole but inside the Arctic Circle, for example, so as to overtake yesterday – and, if we go on long enough, to overtake the day before yesterday. Suppose we speed up this aeroplane in which we are travelling, and continue to speed it up and keep speeding it up. We shall find our day shrinking to twenty-two hours, twenty-one hours, and so on.

A twenty-three-hour day on a liner means merely a slight loss of sleep at night, which is easily remedied by a doze in a deck-chair during the day; but when the day comes down to fifteen or sixteen hours we shall experience a considerable dislocation of our traditional routines, sleep, meals, working time, appointments, and so forth. We shall experience these inconveniences because, measured by heartbeats or by a chronometer, our lives have undergone no sort of acceleration whatever. We are no quicker in the uptake and in our bodily movements. The flicker of outer events goes faster and faster, but our powers of response and intervention do not increase. It becomes a fine question whether Time is getting the better of us or we are getting the better of Time.

We are living more days and getting more experience for the same expenditure of vitality; that sounds like winning against Time. But we are being obliged to alter our routines and adapt ourselves to the new tempo, and that sounds as though Time were the driver.

That sensible person, still clinging to the chronometer as the last word about time, objects. All this is exaggeration. He says that all you need to do is to stop your aeroplane and get out, travel from east to west in moderation, and so return to a normal pace of living. "People always *have* made such small necessary adaptations," etc.

I do not know what the sensible person would say to a suggestion that perhaps we might rearrange the divisions on the face of the chronometer. Our ideas about days and weeks are still very Babylonian. Most people in the world are still sufficiently sedentary to find a practical significance in midday and midnight. So that it is perhaps a shocking suggestion that for an increasing number of interests it might be more convenient to have a common time and a common date throughout the world, a simultaneous Now, and also to have a local time in relation to the sun. You might have a local time and a world time. Suppose we set all our chronometers abreast, and suppose we dealt with an hour double the present one and a ten-day week. The navigator uses Greenwich mean time throughout the world, and for international travel we have already found it more convenient to deal with a day of twenty-four hours and to talk of "Zone Standard Time" instead of a day and night of twelve hours each. For transcontinental flying one is always messing about with one's watch as one passes from one Zone Standard Time to another.

Great multitudes of people at present are trying to follow the war news. They find the utmost difficulty in grasping the sequence of events. The future student of the history of these times will need a variety of calculating machines if he is to realize how event followed event. None of this slights the veracity of the chronometer, but it does suggest that, just as we have had to abandon a gold standard for exchange purposes, although gold is just what it always has been in our scales, so we may have to substitute other units in our time reckoning. We may find ourselves in a world of four or five working shifts a day, we may turn day into night, we may adapt our minds to realize that five o'clock is not the only tea-time in the world nor eight the time to put the children to bed.

Things are not so simple as that sensible person imagines. It would be all very well if we had only the speeding up of east to west travel to consider; but that is merely my approach to the realities I want to discuss. The acceleration is general. We are not only travelling east and meeting our tomorrows prematurely, but also we are travelling west and picking up our yesterdays. It is not only planes that fly faster and faster, so that they can beat the sun at lower and lower parallels of latitude, until at last, when they can go over a thousand miles an hour, they will beat it at the Equator, but it is also a matter of messages that take practically no time at all, loud voices which are heard simultaneously from end to end of the earth, and such contrivances of transmission that the face of a man who was unknown yesterday can become a familiar presence everywhere tomorrow. Events may be taking the same amount of time by chronometer as ever they did, but there is more and more interstitial squeezing.

It is the intervals between events that are dwindling to nothing, and, for any resistance we may put up, that crowding together of events goes on. The Thebaid is no longer a safe hiding-place; the monks of Mount Athos are ousted by anti-aircraft guns; the simple hermit, "the world forgetting, by the world forgot," finds his spring of water poisoned for strategic purposes, he is forced to change clothes with a sniper, and bundled out of his cave in uncongenial regimentals, exposed to incalculable enemy action. There is no escape from this searching and accumulating abundance and simultaneousness of living.

It is a literary convention to speak of all this as the "wild rush" of modern life. It is a wild rush only for those who aspire to omnipotence, omniscience, and omnipresence. I can understand Jupiter's head spinning faster and faster as the pace increases; I can understand headaches in Heaven; and for anyone else who wants Everything. For the greedy there is indeed indigestion; but there always has been indigestion for the greedy. For minds, on the other hand, that are content with a finite realization of their innate possibilities, life is becoming not only immensely fuller but immensely easier than it ever was before.

> Time like an ever-flowing stream
> Bears all its sons away.

But not so fast as it used to do; not until they have got far more out of life than any preceding generation. We live twenty times as fully as our grandfathers. To that extent mankind has already beaten back the ancient aggressor. And that is only the crude statistical opening of the story of our conquest of Time.

"*Tempus edax rerum*," says Ovid, putting the idea in another form; Time the glutton, cramming everything into its maw. Like most of the major Gods up to and including our Christian Almighty, Chronos has been a proleticide. But now the bonds of Prometheus are relaxing.

It did indeed seem to Ovid as though Time had swallowed many things so that they were altogether forgotten. But now

Time swallows with less assurance, looks doubtful, stops eating, and turns green. Not only do events go on record, and keep on record, but Time begins to disgorge. Every year we win back more of the past history of the universe and know its particulars more surely. Our great-great-grandparents were confronted with a silly mythology that reached back to a special creation of the world sixty centuries ago. Now we have forced open the lying jaws of the great devourer, who has not really devoured but only seemed to devour. He has been unable to swallow the evidence against him. The stars, the rocks, the atoms: their past becomes our present. Here they are round about me in my library and in restorations and documentations that every day become more accessible to me, and more easily and swiftly accessible.

Just as man in this present phase is conquering the spaces that keep him apart from his fellow men, sinking his probes deeper into the earth and mounting to the stratosphere; just as he shakes off the imperative to toil, making the once wasted energy of matter his slave; just as he can now quite easily grow all the food in all the variety he needs; so now also he begins to free himself from unending hurry, the relentless timetable, the panting pursuit of occasions that once lost are lost for ever. He can contrive urban and suburban roads that will carry him at his ease to airport, seaport, or wherever he wants to go – he need not follow them for they will carry him; before very long he will be able to summon everything there is to be seen, every machine, every show, every living thing, every masterpiece and movement, in its utmost vitality and in any detail, to his study table; he can hear all the music in the world, and, if he wants to do so, all life's edifying discords. All this he will be able to do *whenever he chooses to do it as a species*. For all this we have chapter and verse.

The experiments have been made; the samples pass muster. I have cited them in my *World Brain* and in *Science and the World Mind,* and I will not encumber my argument by a recapitulation here.[1]

These are man's present possibilities; and without haste and without delay he can complete his material conquest. He will soon be able to talk to anyone anywhere, be within help of everyone, and laugh at the tides and seasons that once chased his hunted heartbeats round the year.

All that concerns our relations to real time, to objective time. But there is also, a friend reminds me, a different sort of time altogether, and that is subjective time, which is steadily shortening throughout life. He writes:

"What I think is interesting with regard to time and human life is that the spacing of time related to one's inner life is not constant, but follows a sort of inverse logarithmic curve. The period of one's life, for instance, between the ages of twelve and sixteen is enormous – yet only four years. What a flash in one's life is the period between the ages of say sixty to sixty-four! I think that gives an example of the unevenness of the flow of time related to one's consciousness, and that is the only way time means anything to us. Time when unconscious is meaningless."

Which is all very well until we reach his last two sentences, when he seems to me to fall over the bright edge of his own idea. But if we read this with the suggestion of an expanding Now in our minds, we get to a conception of the personal life *escaping from the urgency of activity towards a conclusion in contemplation.* The gradual convergence of all lives upon a common existence, when the body is cast aside like a garment and when past and present have a steadily diminishing significance, will be discussed again in "The Religion of the New Man" (Chapter 10).

This assertion of approaching victory for humanity, this ending of the Martyrdom of Man, may sound strange to many of those who are suffering from the acute distress of this present age. But it is precisely because they are still enslaved by Time that they fail to realize that, not in any sort of metaphor but in actuality and reality, past and future are dissolving into an ever-expanding Now.

1 See Appendix I.

CHAPTER TWO

What an Individual Is

In order to develop this statement of reality it is necessary to ask at this point, "What is an individual?" Who are we really talking about when we speak of birth and death? What is this queer enclosure in Time we accept so uncritically as a "lifetime"? Man has been conscious of himself for only a few score thousand generations, and it is not so very surprising that his thought consists largely of rash acceptance and heedless acquiescence.

The aristocratic aloofness of philosophers from rude activities, together with the disposition of the vulgar to avoid philosophy, humbly and respectfully indeed but very completely, as something of no earthly value whatever and susceptible of no practical tests, results in the absurdity of there not being any particular philosophy at all, but a muddle of inconsistent, ineffective, and untried "schools" of philosophy, each sounding its own particular interrogation and leaving things at that. But while the dignified dance of the schools of philosophy, for the most part very elegantly capped and gowned, still goes on, the observational science of natural history which, in the past hundred years, we have rechristened biology, has been throwing an increasingly actinic light upon this common idea of individuality, so that now, to the biological eye at least, it is extremely faded.

The Young Aristotle, who invented his syllogism long before he took to the accumulation of hearsay knowledge (the "Science" of the Old Aristotle), swallowed universal individual separation, just as he swallowed slavery, as a universal necessity. Biological fact shows not only that when we go back in time there *is* a direct continuity of individuals, but that the apparent detachment of the individual is a characteristic only of the higher forms of life. A lobster or an octopus comes into the world, lives its "allotted span," and dies when its time has come, but its individuality is little more than a transitory bodily independence. This is equally true of most vertebrated species until we come to the birds and mammals. The lower grades of living animals and all the vegetable kingdom branch out, split up, remain attached to one another in colonial organisms, and defy any attempt to define an individual. And even as we mount the scale towards our culminating selves the closer we test and the more sceptical we become about these apparently free and independent detachments of time that we speak of as "the lives we live."

The convenient natural disposition of a young human being with a body, including a brain, which has to be controlled, is to regard itself as a simple and complete unity, just as the convenient natural disposition of other individuals is to regard it as that. It has to project itself to itself as something to be steered through a dangerous social world. It has to behave as a creature that will be consistent with itself, and the disposition of its fellows is to judge it by its general behaviour. It ignores completely the manifest discontinuity of its life. It disregards the fact that it can pass over gaps of duration during which it is completely insensible or about which it remembers nothing. It ignores the fact that considerable chunks of experience seem to have slipped out of clear memory and make themselves felt by obscure deflections of behaviour.

A man in his ordinary consciousness isn't "all there," but, unless and until he falls under the investigatory care of a psychoanalyst, he is unaware of it. From the point of view of the average man, he just has his "oddities." Freud, dealing with the mental troubles of a peculiar social section of Vienna, devised a method of attack upon this stirring mass of stuff downstairs, under hatches, as if it were an accumulation of forgotten phases in our apparently continuous lives; Pavlov, attacking it from a more radical and conclusive experimental angle, showed that at no time had this Sub-Conscious been incorporated in a consecutive individuality. There is not that much solidarity in the Ego. To the Behaviourist the mind is a complicated system of reflexes, held together by a body. It is not an originally unified assembly whose factors have lost touch with each other. These factors are, and they always have been, unaware of each other.

The mind as it appeared to Freud may be described as an analysable unity – it went about dropping things and picking them up again; to Pavlov it was a progressive synthesis, gathering itself together but never yet achieving unity.

Neither branch of this modern thinking leaves us with much of our naïve original assumption that our detached lifetime is something with a beginning, a middle, and, it may be, either a definite end or a definite separate endlessness.

CHAPTER THREE

The Fear of Death

This personality of the higher animals, this consciousness of a self which has to comport itself, *is a serviceable synthetic illusion of continuity that holds the individual behaviour together*, but it is neither so lucid nor so persistent as to involve the apprehension of an end.

It is doubtful if any creature below the human level has the faintest idea of death. Fear is a very primordial thing, but it seems to be always a more or less definite dread of some disagreeable or horrible living experience. It is not an "instinct for self-preservation" that makes a reindeer run from a wolf. It is just the horrible wolf that causes it to run. It is a direct response. And it eats not to preserve its life, but for the sake of eating.

Animals will kill or witness the killing of their fellows in manifest ignorance of what killing means. Medieval courts have tried and condemned animals for murder, but whether an animal can commit murder of set intention is a very fine question for an animal psychologist. It may murder as a burglar may murder for robbery, or like a man who "sees red" in a fight. A ferret kills with gusto; but is it for the sake of killing or for the warm delight of bloodshed? And does the cornered rabbit scream at death or at frightfulness? Nature below the human level may be "red in tooth and claw" and still fundamentally innocent of killing.

The idea of death may be a very recent addition to the complexity of life. And this unawareness of death as an event ahead is true of young human beings, and true even of many people for the greater part of their lives. Many of them hardly think of growing old. Their minds turn away from it. The invasion of consciousness by the ideas of death and decay marks an epoch. Then it was Time launched his great offensive against the serenity and continuity of our race, and for forty or fifty centuries carried all before him. Grown-up people were overwhelmed by it, and they poisoned the lives of millions of innocent little children who would never have thought of it before their adolescence, and filled their minds with a new and more pitiless fear.

The tyranny of Time as the Death-Bringer over the opening phase of human thought was monstrous and universal. As the human communities grew in size and complexity, as the development of interdependence and of mutual dominance, service and servitude, progressed, great organizations and buildings appeared, to consolidate and testify to this prevailing anxiety. Religion is woven of many strands – primordial magic, fetish, taboo, various forms of magic necrophagy culminating in recent times in the Mass, excitement by the sight of blood, abject fear of the vindictive sublimated Head Man, phallic perplexities, the practical need for a calendar; but its evolution throughout the ages is chiefly to be traced by the great mausoleum temples it has spread about the earth, and by the traditions and creeds embedded in the human mind. Out of these centuries of Death-worship and Time-assuagement the human mind emerges at long last. It emerges to the dismay and anger of every priesthood and priest-obsessed victim in the world.

The crucifix, that pitiful, hideous symbol of torture, death, and human abasement, commemorating Paul's triumph over the Nazarene he misrepresented and betrayed, fights to sustain our waning fear and abjection in the light of that dawn. And the Church, which exploits our terror, fights, as everyone who has followed the controversies of the past century knows, without dignity or candour, with all the characteristics of a conscious inferior; obstinately, abusively, cunningly, meanly; boycotting and betraying. Whenever the Catholic prevails he censors and burns books; and always the ultimate weapon of his spiritual tyranny is a cultivated and exaggerated horror of death. He holds the keys of heaven and hell. He alone, he insists, has the means of escape. Accept the priest, swallow his mythology and then –

> Oh Grave where is thy victory?
> Oh Death where is thy sting?

But what becomes of him when we realize, as we are doing now very rapidly, that Death has no sting and Time is on the defensive?[1]

---

1 The comment that has just been made on the relation of St Paul to Jesus of Nazareth may be the clearer for a few pages of elucidation. Christians read their Bible very carelessly. Most of them do not read it at all, and even professional Bible-Readers seem to be in the habit of thinking about something else while they read it. But they feel very strongly about it, and the more so nowadays in these dangerous times when you cannot be too careful what magic powers you offend. What I have to say in this Footnote about these two people is based entirely upon reading the Bible. I call no other witnesses at all.

The accounts we have of Jesus and Paul are given to us in four "Gospels," which vary widely in style and contradict each other upon a number of particulars. It is generally regarded as impious to ask why Almighty Providence should have imparted the good news of man's salvation in this slovenly fashion. Jesus, I was taught, was the son of the Virgin Mary and the Holy Ghost. I understand that is the universal Christian teaching. Yet for some fantastic reason the Gospel according to St Matthew begins with the genealogy of Joseph, who seems to have had very little to do with the business. The Gospel of St Luke gives another and different genealogy for this irrelevant putative parent. It is, I suppose, blasphemous to compare the two. So I will not exhort the Devout to make this comparison for themselves if only they will permit me to make it.

A plain account of the origin and relationship of the Holy Ghost should be of far more interest to Christians who have to believe in the doctrine of the Holy Trinity. It is never given. It would have saved endless bloodshed if it had been stated without ambiguity. I may perhaps be permitted to remark that in the opening paragraphs of the Gospel of St John the Holy Ghost has an odd resemblance to the Logos of the Stoics, and that he does not seem to have entered the Trinity until well after the New Testament was written. St Athanasius, or some other person using his name, found the information supplied by the inspired word of God so unsatisfactory that he had to supplement it by his own well-known Creed, "which except a man believe faithfully he cannot be saved."

None of these Gospels, it is now generally admitted, were in existence until a number of years after the Crucifixion. They seem to have been compiled from earlier documents and patched up independently. Their defects and patchings and obvious interpolations give them a sort of independent authenticity. No cunning fellows would have produced this naïve jumble of muddle-witted stuff. The first three Gospels do convince me that there must have been a Jesus of Nazareth who was crucified, that he had a following who believed in him personally and intensely and who, after waiting some time for his return, set themselves to put on record all they could remember about him before it was completely forgotten. Their minds were pathetically and unintelligently loyal to their vanished leader. They stuck in all sorts of things to enhance his credit – those irrelevant Joseph genealogies, for example; they polished up his wonders and miracles. It was in the current tradition that anyone with a "message" to mankind should justify his pretensions by some high-class conjuring, and the Gospel writers thought with their times. John the Evangelist may have produced his Gospel later than the others; he was evidently under the sway of Paul's theology and he touched up the story more impressively, making Jesus the Christ beyond all question.

For the amplification of such questions the reader is referred to the *Bible Encyclopaedia*. Our interest here is in the real Jesus who appears dimly but confusingly through these tattered, worn, and maltreated documents. It is plain that he belonged to an old order of intellectual life, in which wisdom was transmitted by word of mouth. He was a precocious child, and when he was twelve years old he slipped away from his parents when they visited Jerusalem, and was found after an anxious search sitting in the midst of the doctors, both hearing them and asking them questions.

For some years after that his story is obscure; he grew in wisdom and stature, but it did not occur to him that it would be a real economy of effort and prevent endless misunderstandings to write down the "message" with which he was beginning to feel he was entrusted. This strengthens our realization that he conceived of his message as an oral message because he could not read or write. In the entire Gospel literature there is only one statement that even seems to contradict this. When he defended the woman taken in adultery, he bent down and wrote with his finger in the sand. What did he write? Surely from the Christian's point of view it was the most precious piece of writing in the world! – if it was anything but a mere scribble in the sand.

For some years after that visit to Jerusalem, the Gospel narrative fails us altogether. It was a time of profound political unrest and it is natural to infer that Jesus was talking politics and developing his ideas in Nazareth. There was much talk of the coming of a Messiah. John the Baptist was stirring up people, urging them to repentance and expectation. The Messiah was at hand. John denounced the uxorious follies of Herod the tetrarch of Galilee, and was put in prison. Jerusalem, however, was in the hands of a Roman garrison. The King Herod of the Nativity, who had given an air of independence to Judaea, had long since passed away and Caesar had strengthened his grip. Jesus emerges in the Gospel story, a figure of power and inspiration, preaching the coming of the Kingdom of Heaven. Plainly it is an earthly kingdom he has in view. Read the Gospels. He is a figure of righteous anger. He curses the barren fig tree. His march upon Jerusalem was a militant one. He created a riot in the temple, driving out the merchants and overturning the tables of the money-changers. The people rose with him, so that he filled the authorities with dismay. He went into the garden of Gethsemane, and his following was armed with swords and staves; Simon Peter had a sword with which he cut off the ear of the high priest's servant. But the mass of the people had veered from their first enthusiasm, and it became clear that the insurrection was ending in defeat.

The story of the trial and execution is all the more convincingly a story of real events because of the manifest discrepancies – the behaviour of the thieves, for example.

At the end upon the cross came a bitter cry: "My God, my God, why hast Thou forsaken me?", and then a last cry of despair.

At least, so say Matthew and Mark; but Luke has substituted other words less perplexing for the faithful, and St John, well saturated with Pauline doctrine, gives a third alternative to the Believer and makes him say "*It is finished,*" and bow his head and give up the ghost. There are three sets of last words, each one flatly contradicts the others and makes them impossible, and no Bible-Reader ever seems to have observed that. We seem to have every stage in this fourfold record between the tragic truth, the more edifying concoction, and the purely doctrinal falsification. St John is very emphatic, almost over-emphatic, that he was an eye-witness of the Crucifixion. It is almost as if he knew of the other versions of the story and wanted to write them down.

Plainly the very real and convincing personal story that emerges from the stained and mutilated Gospel records, with their foolish glosses of "which was done that the Scriptures might be fulfilled," and their strenuous doctrinal intensifications, is one of a social and political revolt and defeat.

It is to be noted that Jesus was an exceptionally weak man in many ways. The Devout talk incredible nonsense about his sufferings upon the cross. But he suffered far less than the two thieves who were crucified with him; he was dead in six hours; it was unnecessary to break his legs. The two thieves had their legs smashed to hasten their deaths, because the morrow was some particularly important sort of Sabbath for the Jews on which it would be improper to let men die. But he did not even have his legs broken.

Crucifixion, it has to be noted, was a death involving hardly any bloodshed. The efforts made by the Gospel copyists and improvers to bring a little blood into the story, and so fit in the growing myth of Christianity with the blood bath of Mithraism, are manifest and pitiful. All this is plain and clear to any intelligent person who brings a modicum of common sense to the reading of the New Testament.

On these considerations we base our statement that later on Saul of Tarsus exploited the still very considerable Nazarene movement for his own elaborate theological inventions. He did not know Jesus, but he knew that there was still a widespread distressful feeling that this valiant leader would return. The imaginative history of mankind is full of these sleeping heroes who will come back to us. Interwoven with Saul's sense of a large possible following were the theories of the seed-time sacrifice of a human being that haunted so much of the old-world religions and still survive in the Mass. (It would, by the by, be a useful exercise for the habitual Bible-Reader to find how often this cardinal Christian ceremony, so important that one cannot die without it, is mentioned in the Gospels.) But to Saul, who was a hot persecutor of revolutionaries, it came as a brilliant stroke on the high road to Damascus that Jesus could be represented as that annual sacrificial seed-time king. And disregarding that human effort at a revolution for righteousness that is still so traceable through the Gospel jumble, and which he was temperamentally incapable of understanding, he made that pitiful failure and execution the sole significant fact in the career of Jesus.

And here again the New Testament is our evidence for another contrast between Jesus and Paul. Jesus, it has been pointed out, belonged to the ancient tradition of oral teachers. He never wrote anything, and probably could not write. Paul, on the other hand, was a copious able writer. The student of the Scriptures passes from the muted passionate rebellion against the wrongful things in the world that underlies the Gospels to the brilliant speculations in the Epistles, of a worldly intellectual who felt no scruple in nailing Jesus for ever on the cross of his defeat. There is no need to call any other testimony.

CHAPTER FOUR

The Exaggeration of Pain

The dreadfulness of death arises out of our still very immature and confused habits of thought. Young people think that they will know when they are dead. They cannot imagine themselves dead; they imagine death as a new sort of living. They conceive death rather as a conscious paralysis, a stiff awareness of impotence. But no man will ever know that he is dead. You may know you are dying, but that is because you are still alive.

Because of the queer beliefs that have arisen out of the facts of dreaming there is an idea that we are made up of two (or three) elements, body and soul, body, soul, and spirit (people are not very clear about these), and there is an idea of "dissolution," a wrenching apart of these two (or three) constituents. It is painful, one gathers, and we have such phrases as "the last agony" to kindle our apprehensions. The vile body remains. Soul and/or spirit depart in some direction not specified. It used to be radially from the earth's surface, but that seems to be given up. Priests discourage inquiries, and are apt to become evasive and irritable when they are made.

There is no Christian anthology upon this matter. If ever one is done, it will probably be done by the Rationalist Press Association. The Christians dare not do it.

And yet that disembodied life after the body ought to be of the intensest interest to everyone who subscribes to it. Consciousness, one gathers, goes off with the detached material, but how it carries on without a periodically aerated brain is wisely hidden from us. We do not know whether it stays behind in space, or whether the believer considers it obeys some obscure super-gravitation that still anchors it to a definite locality on our spinning planet in its headlong rush through space. There is not an atom of proof or probability of any such detachment. In real life any interruption of the blood-flow to the brain produces immediate unconsciousness. This dissolution nonsense has been handed down to us from the childhood of the world. The fact is that we die, and that before we die sensation fades.

Not only is this true, but it is equally true that there are very fixed limits to human and animal suffering. As a boy I heard a shrill missioner in Portsmouth Cathedral trying to impress us with the tortures of the damned. Every moment the damned individual was to experience all the pain that has ever been on earth and more also. Even at fourteen it was impossible not to feel that this Christian God of Hell was an utterly detestable maniac, demanding hate and defiance at whatever cost. I am told that now things are very different, even in Catholic teaching, and that the God of Hell and Salvation has been making himself better understood even in the most authoritative circles. Hell, it seems, has been closed for repairs and reconsideration, and may never be reopened. That is good news for countless scared little children – if it gets to them. However that may be, that tenoring fear-monger was proclaiming an utter absurdity. There are limits to pain, and they are being contracted.

Nature knows nothing of an unchanging pain. If a stimulus is applied to a pain organ, and then continues to act on it with constant intensity, the pain organ responds by sending up a volley of impulses of brief duration, then becomes silent. To keep a pain organ actively sending impulses, the stimulus must continually increase. So you cannot keep an absolutely constant pain sensation for long. The instant spasm that produces an immediate recoil is no pain at all. The twinge must throb and return.

Nature is neither considerate nor inconsiderate; but she is indifferent and devoid of any divine malignity. She permits the most disgusting-looking events in the way of parasitism, particularly among the smaller fry – grubs that infest other grubs and devour them slowly alive; but that either eater or victim has any distress or satisfaction in the process is improbable. You have to imagine the victim is a nicely brought up child of the more sensitive classes before you can enjoy the full horror of the situation. It is questionable whether the fly upon the window-pane or the fly in the spider's web is really the fine-minded distressed little gentleman we assume it to be. There may not be a shadow of pain in all that drama. A number of genteel Sadists will resent being robbed of the gratification of pity; but it may be so. Even man can have his liver infested with liver fluke, or entertain enormous tapeworms, without conscious distress.

On the other hand, our nerves are so ill-constructed and so easily disorganized that growing pains, the aches of neuralgia, lumbago, migraine, and so forth occur, and have caused and still cause a vast amount of entirely purposeless suffering. But not so much as they did. They can all be anticipated and subdued. Toothache, which is probably the most distressful pain that man can stand, is now quite unnecessary. It happens because our lives are still badly managed. The pains of child-bearing, again, are quite unnecessary. The sufferers cannot get the proper stuff and attention, but the stuff and treatment exist and are known. It is disgusting humbug to say that only God can give health and contentment. In the world of inequality which is now collapsing the prosperous are far healthier and happier than the poor. They live longer and they fear less. Properly organized upbringing, nourishment, doctoring, and properly educated behaviour can banish nearly all suffering from a human life.

We are living our way out from a maximum exaggeration of the Ego. We are living in the close of a phase wherein it has been unnatural to think of ourselves as non-existent. In the opening phase of individual life we are so unaware of all that we are unaware of, of its gaps and losses, that it is natural to believe ourselves continuous and immortal. Even when we grasp the full implications of Pavlov's conditioned reflexes, we are still disposed to assume that presently they will extend and build themselves together and become one unified self. But why should they? What necessity is there in that?

Neither from what we know of the inner or outer imperatives of the animal we are is there any reason for that expectation.

CHAPTER FIVE

The Fear of the Dead

What goes on after the death of a human body?

Nothing in that human body except decay, but much outside it. The behaviour of our species towards these dead bodies betrays a confusion of fear, propitiation, and a greedy appropriation of the belongings they were no longer able to defend and control. The survivors grabbed in the daylight and then were horrified by vivid dreams. The dead haunted them. Down the ages the dead seemed to be demanding respectful sepulchre and commemoration, and ever and again it chanced that a bad harvest or a pestilence enforced the menace of their immortality. Or maybe they haunted because they had suffered wrong; and then there was a witch-smelling to discover and avenge the hitherto unsuspected crime.

After a time fresher ghosts ousted or were identified with the fading nightmares of an earlier generation. A certain number of loving spirits, unable to believe their dead were dead, stayed like dogs by the dear body, refusing to be comforted, and kept its memory green – but after a while perhaps not quite so green – until they too died or were distracted and partially forgot.

The desire to keep the dear body fought against decay. The silent, motionless thing was dried and embalmed; it was put in a mummy case that had a portrait face; it was covered up in a monument upon which the figure of the deceased was presented, larger than life. The bereaved carried about miniatures and rings and relics. Many people do indeed treasure so vivid a memory of the sayings and doings of those whose love they have learnt to believe in that it may last to the very end of the mourning life. The wishes and preferences of the trusted companion-lover may exercise a greater power beyond the grave than in actual life. It is not given to everyone to love, but to love self-forgetfully can make one part of another life, so that one lives on in that.

This is the way of sensitive minds; but the more common reaction to the dead is and always has been fear, the fear of the haunting spirit, and the barrows and pyramids and temples and monuments that spread like a pimply rash over every region of human habitation during those centuries of Time's tyranny were mostly propitiatory offerings to induce a dreaded Personality to settle down and keep down – and out of the way. They were, in fact, Dead Weights.

The idea found a response in every domineering Personality. A man could not take his appropriations with him, but he could at least impose upon his eager heirs the duty of pickling and preserving his invincible self. He could endow his memory. They were only too anxious to put an adequate Dead Weight upon his body, and he was only too anxious that his name should live for ever. The monument was the compromise. And since the departed are apt to be strangely silent it was an obvious opportunity for others to cherish and interpret the will of the mighty dead. They would transfer their hatreds to him, and have dreams and revelations of his usually vindictive purpose towards the disrespectful.

So the prophet, the priest, and the stonemason worked together at making life Monumental. Everywhere the memorials and cemeteries cluster. The economic development of China was much delayed by the fact that hardly anywhere could you dig without disturbing an ancestor. Great and little conspired to sustain the accumulation of anniversaries and names. The burial society assisted in this salvage from the waves of time. At the end of his inconspicuous life John Smith still held on beneath his humble cross and his footstone. "John Smith of this parish, aged fifty-nine, died September 26, 1875, in hope of a glorious resurrection" said:

"*Me*, please!"

Today John Smith's little memento (1875 vintage) is likely to be rather mossy and neglected. Maybe some faithful hand still keeps it tidy and cuts the grass; but it has not the freshly renewed flowers of the later mounds. And if presently we wander into the older corners of a typical country churchyard, where the John Smiths of the eighteenth century are tucked away, all these signs of personal remembrance vanish, and, it may be, some of the stones have their names obliterated and stand askew. And from that we pass to those urbanized cemeteries where all the old stones are piled against the wall, and only the railed-off dignity of the richer sort remains. And so, more and more weakly as he recedes into the past, John Smith keeps up his pitiful "Remember *Me!*" until he is lost for ever.

But the Kings and the Crusaders hold out far down the centuries, the saints and deified ones. They mingle at last with creatures of the imagination.

I will but glance at Mrs Gummidge making the dead a weapon of domestic tyranny.

Yet always in secular literature there are traces of a certain rebellion in the mind against this too oppressive pickling and preservation of personalities. There have always been mockers of the sepulchral tradition, and they have multiplied. The medical students' second-hand skeleton is crowned with a bowler hat and given a clay pipe between its teeth...

Human beings are transitory. The mind rebels naturally and very readily against the tyranny of dead Persons.

CHAPTER SIX

The Fundamental Question of Philosophy:
the One and the Many

Human beings die and pass. And in this present period of world warfare, in which there is a storm of strange and violent killing, in which, indeed, everyone is involved and people die far more significantly than they have ever died before, it is nevertheless natural for many to ask with Macbeth whether life is anything more than a tale told by an idiot, full of sound and fury, signifying nothing? Has it any quality of will and purpose in it at all?...

Is that question, in the light of modern realizations, badly put? Did Macbeth–Shakespeare phrase the case in such a form that an unrealized alternative slipped through his fingers? It may be that it is true or untrue, subject to certain stipulations. Our decision may be dependent on the assumption that the only thing we can call life is the individual life. So Macbeth puts it. When he calls life a "brief candle," talks of "a poor player who struts and frets his hour upon the stage and then is heard no more," he is manifestly thinking of the individual life, the "yesterdays" of that neglected corner of the churchyard. Shakespeare–Macbeth disavows immortality and is frankly Atheist, and that is as far as the most sceptical criticism of life could go during the Elizabethan age.

But there is another life far greater than the individual life, of which we are growing aware, which is not so easily dismissed – the life of the species as a whole. The long obsession of the human mind by the false assumptions of its individual separation, which crystallized in the Aristotelian logic, is lifting. The human intelligence has, indeed, swayed right across the reality of the matter to another extreme, to the idea of the complete subordination of the individual to some larger Being, the Flock, the Gang, the Party, the Community or State or Species – it is put very variously, but the essential idea is the complete subordination of the individual life.

Now it is fairly evident to those with a sound contemporary knowledge of ecology, that branch of biology which deals with the relations of a species to its environment, that Nature does not care a damn either way whether the individual survive or perish. She is the extreme neutral. But the same fact is equally true of species. The process is a little longer and a little larger, and that is all. The individual must fit into the Time process or die; but so must species and genera and orders and classes, and maybe in the long run all the kingdoms of life. Nature neither hastens nor delays. Survival is the affair of the surviving being.

But between individual and species there is a relationship of a sort that does not exist between the diverse species that have a generic relationship to each other. The individual and the species return into one another in a fashion that has been the chief concern of philosophical inquiry since philosophy began – the relation of the One to the Many. Philosophy has fluctuated between the extremes of the Realist school, which insisted that the One, the Common Frame, the General Term, the Idea, the Ideal was supreme, and the Nominalists, who insisted that individuals were all-important and the General Term a mere name thrown about them, a bag to carry them about in the mind. The conflict of these two trends of thought led to Neo-Nominalism, which realized that classifications were neither arbitrary nor fundamentally true, and so released that persistent reference of statement to observationally or experimentally determined fact, which nowadays is called Science. But the most difficult thing for the human mind to do is to balance, and while Natural Philosophy (or, as people say nowadays, "scientific inquiry") went on along the line of science, in the expectation that a great free correlation of individual effort would ensue without further effort as the collective truth, large sections of human thought fell back upon the old Realism, subdued the individual to insignificance, and exaggerated the specific Whole. The individual, they thought, was a mere pseudopodium, an experiment, an exploration thrust forward and outward to bring in new experience and build up Leviathan.

The world that saw the early meetings of the Royal Society also produced, independently and simultaneously, Hobbes' *Leviathan*. The original frontispiece to that book displayed the monster as a completely unified social organism into which all the estates of the realm were built together and surmounted by the visage of the sceptred and anointed sovereign. From the days of Hobbes to the National Socialist literature and that "philosophy" of "Holism" of which General Smuts is the exponent, there is a vast literature of synthetic Stateism.

Men released from the comprehensive "one body" of the Catholic Church have always been particularly prone to this type of self-surrender. Hobbes, like Hegel, had a timid and subservient character, and it was natural for him to give the collective human beast the image of the social order that evoked in him the fullest sense of "belonging," in security. What he did not grasp was the essential *interdependence* of the individual and the group or species. It may be well to restate that interdependence here.

But first we may be reminded of the practical unity of every species of the higher sexual animals. Every individual has two parents, most have four grandparents, up to eight great-grandparents, and the farther we go back in time the ancestral tree spreads. Most people who have read any embryology will remember that before a fertilized ovum develops farther, the nucleus gives off two small bodies – the first and second polar bodies – and that these bodies are known to remove three-quarters of the inherited tendencies of the maternal parent. The spermatozoon of the male is not the equivalent of the ovum in this matter. It is the equivalent of one-fourth of the ovum's burthen of inheritance. (There is a full account of this in Wells, Huxley, and Wells, *The Science of Life*. That book is published as one whole work, but until recently it was also obtainable in nine separate volumes. The volume entitled "Reproduction" gives all this very clearly and fully.) At each new conception, therefore, half of the fertilized ovum's burthen of inheritance is taken off. There is no reason to suppose this extrusion of the polar bodies is selective. They are just thrown off, as it were. Most probably they are thrown off as associated groups.

Local influences of climate may favour, and obviously do favour, those members in every family who come nearest to certain locally adapted types, but, seeing how human beings drift and migrate, and have drifted and migrated in the past, men more than women, it is manifest that as we recede in time the number of our ancestors must increase.

On the 2, 4, 8, 16 basis we should find that before the time of Queen Elizabeth there could not have been a soul alive who left offspring who was not the ancestor of everyone in England, and, even when we have allowed for much intensive local interbreeding, we should find all our ancestral trees running together and returning into each other long before the days of William the Conqueror. Anybody who came and lived and settled and bred in England before that time, Jew or Negro, Roman, Phoenician, is almost certainly in my genealogy. Pocahontas brought a Red Indian strain into a number of English families, and a storm-driven, shipwrecked junk would have been enough to link America and the Old World. The world of Julius Caesar was ancestral to ours, *en bloc*. The image of spreading trees in the direction of either past or future is a delusion. The real figure should be an interminable network of the species continually returning into itself. One correspondent compares it very aptly to a loofah. And if this is true of the human species, slowest in its reproduction of all breeding animals, it must be far truer of all more rapidly maturing species. *So long as a species can breed within all its mutations and variations, it remains a collective unity.*

It is from a perception of this continuity of the species that the idea of the subordination of the individual to some sort of Leviathan draws its strength. It is a half truth, the converse to the half truth of dogmatic individualism. The reality, as the biological observer is best qualified to realize, lies between.

If a species survives, then it survives only by and through its individuals. It is a great mass of individuals rolling towards that maw of time, and every one of these individuals is contributing to the movement. The species may or may not survive; it will struggle to survive as stoutly as its collective will – or, if you prefer it, the algebraic sum of its wills – to survive determines. So long as it holds out, the life of every individual must contribute some consequence to the struggle. Every individual life without exception changes the species, and its contribution is permanent so long as the species endures.

A species lives and dies in its individuals, and obviously the extinction of a species and the death of its last individual is one and the same event.

The human mind, like everything else in the human make-up, is the outcome of a fluctuating selective process. The modifications of its apparatus of thinking had to have immediate survival value, had to be of immediate practical value, to establish themselves. The apparatus itself was no more a truth-finding apparatus than the snout of a pig.

The practical interests of *Homo* were mainly in the animals about him, and particularly in the most dangerous among them, his fellow men. He had little or no time for admiring the scenery or, until the beginnings of agriculture, star-gazing. He seems to have developed his earliest methods of expression, his gestures, drawings, and early forms of speech almost entirely in relation to these more intimate personal relationships. His first nouns were probably proper names. "Man" was a particular man and "Bear" a particular bear, which he generalized as a baby does when it calls every man "Daddy" and every cat "Pussy." Rudyard Kipling's *Just So Stories* tells how *the* Elephant got his trunk, and every child takes the story that way. It is taken to the Zoo to ride on *the* Elephant; it strokes *the* Cat and hears *the* Dog bark at it. As the human range of interest extended, this way of thinking expanded to include far less classifiable things. All over Britain you find *the* river, the Ouse (Eaux), an individualized flow of water. Man's interest in topography increased with his need for caches and hiding-places, and then rocks and trees and hills and islands became invested with a personality to which they had no right.

Since you could count and mark up on the walls of your cave the beasts or the men you had killed or the nights and days that you had to wait between moons for the love-making, why should you not count and name the mountains in the distance or the paces between your tree and your buried kill?

This problem we have been discussing of the One and the Many applied with precision only to such living creatures as could be thought about as a Type specimen and a series of variations of that Type, but the developing human mind not only classified things that were pseudo-individual, like trees, communities, kingdoms, tongues and peoples, mixtures, substances, rocks (in petrology and mineralogy), but qualities such as colours or moral or sensational values. It was the Neo-Nominalist who finally awakened the human mind to the falsity of its assumption of a universal classifiability, and prepared the way for the realization that even the classes it does accept most readily are separate relative to a section of what Einstein has called the time-space continuum, and that they too lose the separation of their individuals and species and merge into one another when that time space continuum is considered as a whole.

Our yesterdays and tomorrows, our hopes and fears, life and death, and all the sequences of our individual and specific life are no more than a moving picture set in the frame of Relativity. They are the everyday working realities of life. The frame of Relativity explains some perplexing inconsistencies in physical phenomena, but it matters nothing to the actors in the drama of life. It is of no more importance to them than the curvature of the earth is to a billiard marker or a cricketer on his well-rolled pitch.

CHAPTER SEVEN

The Difference Between Mutations
and Mental Adaptation

After these restatements we may now attack the Shakespeare–Macbeth question, and the significance of the spectacle of this killing of body and courage that weighs upon our minds today, with a much better prospect of an answer than before we made them. If all our lives are not a mere chaos of personal tragedies but do make a contribution to the conflict between our species and time, what is the nature of that contribution? We have shown clearly that it is not the Ego, not the detached Personality, that constitutes this contribution. The detached Ego is a working illusion that worked sufficiently well for a time to stampede the dawning human intelligence. Now we can ask, "What is it in the imperfectly assembled mental complex in a human body that does become a contribution to the struggle of the species?"

Up and down the animal series the contribution of any individual to the life of the species is a trial difference. There never was anything quite like that individual before. Nature is not interested in whether it is for the good — whatever we may consider good — or not. Whether the new difference has immediate *survival value* or not is her sole criterion. If it has not, that individual is wiped out and there is an end to it.

A vast majority of mutations are immediately detrimental and so are wiped out at once. The abnormal individual is usually, as we say, a monster. But if the particular mutation is not immediately detrimental the individual survives to reproduce itself, and, in a certain (Mendelian) proportion of its offspring, to reproduce the new variation. Over the area of interbreeding this new variation may spread until it becomes a general characteristic. The value of its contribution to the outlook of the species may vary. It may make for the survival of the species as a whole; it may be detrimental to the survival of the species as a whole; or it may have little or no effect one way or the other.

There is no such exact fitting of animals to their environment as the cruder evolutionary theories of a century ago imagined. The pace of events is too rapid. So the animal kingdom abounds in tolerated absurdities of colour, excrescence, behaviour, adornment.

These relations between the individual and the species hold true over the whole range of individualized animal life up to and including man. Always the individual is a try-out. He emerges from the species with his difference, he put his question "Will this do?," the species takes it on or does not take it on, and anyhow the individual disappears as the specific destiny of the species unfolds.

As we ascend the scale there is an increase in the continuity of life and life. Something more than a bodily difference is inherited by the offspring. The Mesozoic Age, as Professor Julian Huxley has remarked, was one glorious egg-hunt. The distinctive character of the Tertiary mammals was the protection of their eggs and their young. They carried their eggs within themselves, their young were born alive and interesting, if only by reason of the agreeable relief afforded to the mammary glands, and they became not only protective but educational animals. The only things to set beside their parental care were the nidifying and incubating varieties of birds. The parent or parents remain with the young, and the young not only imitate, but, as we rise in the scale of brain development, are shown things and reproved. The young kitten, for example, receives elementary sanitary instruction and a certain initiation in mousing.

But, until we turn to the Primates, the educational factor in life remains a minor extension of inheritance. With them it becomes very swiftly of supreme importance. The Primates branched off very early from the Eutherian stem upon a line of their own at once primitive and original. The *Hominidae* appear, already handing on to their offspring their inventions, their discoveries, and their working misconceptions to a degree relatively immense. Facility of intercommunication made larger and larger social groups possible, biologically more efficient and so unavoidable, and association and co-operation were forced at a rapid pace (rapid in terms of biological time) upon a primarily solitary and ego-centred species. We do not know whether Man's earlier communications with his fellows were made chiefly by sign or chiefly by sound. Sir Richard Paget thinks that explicit human intercourse began with gesture, and that men were even drawing the shapes of things in their caves and shelters before the grunting that accompanied their gesticulations developed into articulate speech.

One repeats these facts to the well-informed apologetically, but it makes our subsequent discussion clearer to recall them with precision at this point.

In this way we approach the developing human society and man adjusting himself to conduct himself. The new possibility of accepting these mental adjustments was so much more rapid in its operation than the older method of physical survival and physical death that it may even have retarded the development of what one may call the fundamental man. The fundamental man, the sub-human man, the man who was allowed to grow up without much social interference, has not been so much evolved into the man of today as overgrown. His genetic character is masked. Real mutations – that is to say, irreversible changes in the germ plasm – have been outpaced by his tremendous mental advances and fluctuations. Which, be it noted, *can* be reversed.

Natural man has, in fact, been replaced in current human society by an *artifact*, a social man manufactured out of the primordial man ape. His nails are clipped, his hair is cut, his natural dirt is washed off, and he is protected by the artificial integument of clothing from climatic stresses; his food is cooked, conserved, and prepared; he lives in caves, huts, and houses; he dies of exposure to the open air in all but a few localities; and his inherent behaviour is checked and moulded by a complex system of education, law, and intimidation.

His sex life has undergone astounding revision and development. Instead of an annual heat, the Primates have a primordial lunar cycle less evident but quite traceable in the male. The innate disposition to maternity at the earliest possible age has been restrained by the imperative needs of the social group; women have grown in stature and freedom until many are, and most can be, as adult as their brothers. In art, literature, scientific research, skilled and responsible work of all sorts, the distinction of sex is fading out. Feminine irresponsibility fades, and, under the stresses of totalitarian war, women, in Russia particularly, fight instead of being fought for. The legislator and the writer deplore the legacy of the more natural past that gives us "he" and "she" for our only personal pronouns. How readily they would say "Man" without distinction of sex, and "he" for "her," if it were not for the natural lingering objection felt by the militant suffragette. Man remains temperamentally heterosexual; but he "mates" now when he used to "take a wife" – or husband – and his sexual life, which once dominated his conceptions of morality and social order, retires now, except in the case of princes and movie stars, into a decent privacy.

The past sixty years has seen this elaborate *artifact*, civilized man, pass, and not for the first time in history, through the completest revolution in his sexual ideas and practices, a process of reversal from a morbid suppression to frankness and lucidity, with a collateral wild outbreak of sexual perversion, exaggeration, and shameless avowal. Things may settle down to a realization that aggressive sex exhibitionism is an intolerable bore in men and women alike, and the rake and the vamp, the gigolo, the painted old girl of sixty, and the knowing self-confidence of sweet rather than innocent seventeen, will fade out in an energetic, busy, affectionate, but personally undemonstrative world. This is not a treatise upon sexual ethics, but attention has to be drawn here to the swift, limitless plasticity of man to his own suggestions, imaginations, and impulses, and sex gives the best instances of that. Human nature cannot be changed, say the mandarins, and if they will only concede that its chief characteristic is an infinite adaptability they are right.

Man is still as curious and experimental as his cousins the monkeys, and he displays something like a passion for self-mutilation. From circumcision to tattooing there is hardly a projecting or exposed part of his body that he has not messed about with at some time or other. And what has been done in the way of experiment to his visible body is as nothing compared with the proliferation of memory and idea-forms in his skull and the accumulation of his social tradition about them.

This tremendous growth of mental artificialities which has accompanied man's rush to material ascendency on this planet has followed a line analogous but not strictly parallel to actual mutations of the germ plasm. Most animal mutations have had the value, a very real value, of negative experiments. They established a limitation. "This won't do. No way here." That is the rôle of the "failure." But since there is no record kept of the mutational experiment, it may recur over and over again. And so practically has it been with these mental innovations.

A few mutations have had direct individual and specific survival value, and spread a new acquisition for good or evil over the destiny of the whole species. There have been similar mental developments of great prepotency, which have become almost universal habits of thought.

But a considerable number of notions in the case of the swiftly prolific human mind were in the nature of superfluous and unnecessary additions to the heritage, comparable to the horns, excrescences, excessive size, prolonged individual life, in many once ascendent orders now extinct. In a phase of adversity such gratuitous elaborations may become an encumbering obstacle to adaptation. Nature betrays no aesthetic preferences; glory, tragedy, or burlesque are all alike to her. There is little dignity in natural history. Most palaeontology is burlesque, a demonstration of the absurd. Possibilities can be tried out in that way just as well as in any other way.

But in the case of Man this proliferation of the superfluous has been almost entirely in our genetic mental superstructure. It is only in recent years that we have begun to realize what a vast jumble of hasty unserviceable notions, beliefs, and dogmas the wisdom of our fathers amounts to, and how enormously they dominate our world. There are dinosaurs and dinotheria in thought, and our mental spectacle today is closely analagous to the condition of the world when the Mesozoic Age came to an end. That was a genetic delirium, and our mental life today, with its Churches and Powers, its Race delusions and its hegemonies, its American Way of Life and its Dictatorship of the Proletariat, and so forth and so on, is a sort of grotesque by Walt Disney, as fantastically unreal.

Through the past thousand centuries of his headlong conquest of our planet Man has been presuming upon his good fortune and sustaining no close touch with reality. But to be out of touch with reality when it involves danger to oneself and others is insanity. By that definition, Man has in fact been drifting towards insanity, and is now manifestly as a species insane.

The human problem therefore resolves itself into the question of how far this insanity has gone too far to be cured – in which case, so far as mankind goes, Macbeth–Shakespeare is right – or whether *Homo sapiens* may be able to learn from his grim experience at the present time and recover his grip upon his mind. Even now, shock may be rousing him from his mental holiday in the preposterous.

CHAPTER EIGHT

The Individual Life is Not a Tale
"Told by an Idiot"

Before I can go on with that heartening series of "Conquest" books, to which I referred in Chapter One, it is plain that I have first to deal with a possible Conquest of Mental Sanity, without which the Conquest of Power, the Conquest of Hunger, the Conquest of Distance, etc., etc., recede into the remoteness of hopeless aspiration.

(So far as man is concerned.)

But it is too often overlooked that while life goes on – and there is no reason whatever to conclude either that it will or will not go on – *every living thing, every living particle, and not only that but every non-living thing outside life, must necessarily be contributing to the career of life in space and time.* The dinosaur has vanished, yes, but it is plausible to assert that without the pursuing egg-hunting dinosaur there would have been no sufficient survival value in the mutations that made the mammals viviparous and both birds and mammals protective of their progeny. The stars are remote, but there is much plausibility in the theory that an astronomical shock spun the plane of the ecliptic away from the plane of the equator, split the equable

Mesozoic year into the four seasons, and so set a premium of survival value upon fur and feather, with their wider range, their need for seasonal adaptability, and the enlarging brains that contributed to that adaptability. But so far these are mere theorizings. No alteration of the plane of the ecliptic will account for the extinction of the marine icthyosaurus, nor is there any satisfactory reason for such facts in the history of life as the disappearance of the ammonites at the apex and end of the Mesozoic. Our knowledge of biological evolution is still only an outline studded with such riddles. But it remains true that while life goes on *everything* contributes to its progressive elaboration. It possesses and uses the universe. And as it goes on it becomes more and more aware of the universe in which it exists. The universe is not known of except by and through life.

We do not know what set life going in time, nor what, if anything, there may be or not be outside its story. Our mental apparatus, so far, is entirely unable to frame any sort of question about that, and much less can it devise any answers. Anything not in the time-space continuum is inconceivable to us. It transcends our powers.

Subtle minds try out analogies that may release us from that acquiescence in the inexpressible with which we have still to be content. Until a few centuries ago man lived on a flat earth under the dome of the sky. Now our existing universe can be best represented as a three-dimensional framework travelling along a fourth dimension in something very nearly approaching a right line – there are slight deviations – at the speed of light. We do not know any reason whatever why that pace should be constant, and we cannot conceive what would happen if there was a direction.

Such a change of direction is at least arguably possible, but inconceivable. We stop at this point – at the end of our wits. [1]

But there is some ground for supposing that our human intelligence is sufficiently in touch with the transcendental to make an increase in its understanding not altogether unhopeful. The transcendental is not chaotic; it is not the story of that Macbeth–Shakespeare idiot; it may be indifferent, but it seems to play fair upon some vaster system of its own. It honours many of our inferences; it confirms our reasoned prophecies. Astronomy is a long record of successful induction; recently the band of the chromosomes made visible at Pasadena has confirmed their theoretical association; and there are other convincing instances. The scientific investigator is guided throughout by his confidence that the unknown is reasonable, and every advance he makes towards it rewards his confidence. Macbeth–Shakespeare was unaware of the contemporary germination of modern science. His mental basis was beyond comparison narrower than ours.

These considerations bring us back to the possibility that if it is through the human mind that life has got to this level of understanding, it may be that the (quite possibly endless) process of life may go on through the progressive development of the human mind. If that is so, we human beings are, each and all of us, contributing, not as accessories but upon the main line, to the awakening of consciousness in the universe. We contribute and pass.

A man clinging to his personal immortality clings to something that changes and vanishes even while he still lives.

Few of us recall the incidents of our early lives, and none of the priests have ever told us what it is that is worth while keeping for ever in a Personality; even the ingenious St Paul dodges and prevaricates about that after-life.

"There are," says he, "celestial bodies and bodies terrestrial... The first man is of the earth earthy, *the second man is the Lord from Heaven...* Flesh and blood cannot inherit the Kingdom of God, neither doth corruption inherit incorruption. Behold I show you a mystery; we shall not all sleep, but we shall all be changed in a moment, in the twinkling of an eye."

To which the only possible answer is, "Thank you very much for your explanation. If only it explained! It seems to mean something to you, most excellent Paul, and you have an air of getting to the bottom of things; it may have meant something to you in that Hellenic phraseology in which you learnt your Judaism at the feet of Gamaliel, a phraseology you yourself use so glibly; even that phrase just quoted, about the 'second man,' may cover some suggestion of that reabsorption of the Individual which we are discussing now; but bless you, Bottom of Christianity, you have been translated now – to nothing..."

Whatever the orthodox Christian view of personal survival may be, there can be no disputing that the impersonal achievement of a man does live on without any sort of ambiguity. The value of his conscious individual contribution resides, as people say, *in the thing that takes him out of himself.* As that remote illiterate teacher out of whose lost Acts and Sayings the Gospels were concocted, and whose prestige was exploited by the energetic, copious, devious, and, as he confesses (Acts VIII and IX e.g.), bloodthirsty Paul, said: "He who would find his life shall lose it, and he who would lose his life shall find it."

That at any rate seems to come as near to the truth about the individual life as we are likely to get for a long time. Much of the life and teaching of Jesus of Nazareth remains hidden from us under the vulgar accumulations of Christianity. He may have been one person, or he may have been the nominal focus of an accretion of traditions of unequal value and profundity. But he does seem to embody and stand for a considerable amount of undocumented deep thinking and meditation. Most of it has vanished into the maw of *Tempus Edax*; but who knows yet what vestiges may survive in out-of-the-way corners of the world, in language and in men's minds?

Some day, on our counter-attack on Time, we may even recover and restore that submerged mass of fine thinking and teaching altogether. We may find that in the light of modern science and philosophy we are restating more effectively ideas that have been troubling men's minds and seeking expression for a couple of thousand years or more.

1 This assertion will be difficult to those unaccustomed to four-dimensional speculations, and so, in order not to make too big an explanatory digression here, I have written a long summary of the essential ideas and facts and put it in an appendix at the end. It deals not only with the significance of Relativity, but also with the very wide range we find among human beings in their capacity for disengaging themselves from old habits of thought and readapting themselves to new ideas.

CHAPTER NINE

After-Man

And now we can come to the question of this present storm of killing, strain, and fear, and ask whether it is the culminating phase before the collapse and extinction of the human race, or whether the human mind will prove itself sufficiently adaptable to survive and enter upon a new and greater phase in its career.

I am convinced that the species we call so prematurely *Homo sapiens* is bound to extinguish itself unless it now sets about adapting itself at a great rate to the stresses it has brought down upon itself. But if it does that, then it will become a new species of self-conscious animal. It seems improbable, though it is not impossible, that it will cease to interbreed freely, and so long as it does not do that it will not split into two or more divergent species; it will evolve *en bloc*.

What can we state with assurance about what is really happening to humanity now, and how far can we anticipate this After-Man, our bodily and mental offspring, of whom this present time is the Advent?

Much of what is happening in the world now is hideous, dismaying, cruel, and shameful; it is a wild storm of elimination, yet nevertheless it is not a biological catastrophe. The physical health of the animal man, though his powers of resistance must be greatly impaired for a time by the almost universal shortage of food in the eastern hemisphere, is still sound and could easily be restored. There has been no pestilence (December, 1941). Scientific control has achieved a conquest of infection – since 1918. If typhus breaks loose, it will be a breaking loose through the killing and displacement of the medical men who know now precisely how to control it. It can be, and probably will be, contained by sanitary cordons. Then such things as the evacuation of urban slums, for example, have led to a considerable cleansing and invigoration of the depressed elements in (e.g.) the British population. There has also been a steady increase in productivity, through a more intelligent economy of effort.

It is easy to exaggerate the mortality due either to pestilence or warfare. We are told that twenty-three million people were killed by the influenza epidemic of 1918. But most of those twenty-three millions were old people, weak people, feeble children who had to die somehow. They fell to the influenza. If there had been no influenza they would have died very shortly in some other fashion. War casualties in the past have been mostly the premature deaths of young men, but a considerable proportion of the present holocaust has meant the miscellaneous killing of defenceless people, already ear-marked for an early death. It is difficult yet to get the exact figures, but the expectation of life today for a human being at the age of eighteen, even in the belligerent countries, must still be very much greater than it was in the time of Queen Elizabeth. All this puts immediate biological catastrophe out of the question. Man is not vanishing.

Equally difficult is it to estimate the amount of injury or stimulation to the human mind that is occurring now. The probability seems to be that the injury is mostly of the quality of insults and vicious assaults – book-burning, the internment and murder of particular writers and teachers, the restriction of publication, and so forth – and that the general mental stimulation is relatively immense. Books are harder to get, but they are read more abundantly. There is much killing, but it is killing of a sort to make people think. A greater and greater proportion of the deaths that are occurring is immediately relevant to the great issues of the time, and their relevance is better understood. Modern economic necessity has evoked communities that read and write from the top to the bottom of any social scale that survives. Intellectually the classless society has arrived. Thought is vulgarized and dispersed, and so it becomes the seed of new volumes of thought. It is no longer confined to rare men in studies and libraries. It has never flowed so broadly as it does now. It is like the Nile in flood, restoring the soil.

In this fashion the new world of the After-Man dawns upon the face of the waters. The dissolution of the old bigotries and intolerances, that seemed invincibly rigid a score of years ago, goes on more and more conspicuously. The chief quasi-mutations in this world of ideas, the new realizations and convictions, that have occurred during that period, are still confined to a small proportion of mankind, but they are establishing themselves in that minority with a consistency and a contagious assurance that make their ultimate dominance throughout the entire species inevitable. Their light is reflected ahead of their advance. Everywhere their influence is perceptible in the utterances and anticipations of the imperfectly informed. The old priestly religious organizations begin to hedge and qualify, the Vatican echoes the New World on Earth note, and the gangster war-makers, who put a cutting edge upon the convulsions of our time, adopt the phrasing at least of a New World Order. Reluctantly, resisting at every step, resisting at the price of an enormous wastage of lives, these vast dissolving organizations, as their younger minds replace their elders, are borne towards the plain common sense of the human situation, towards the threefold need for a common fundamental world law, for a federal conservation of the earth's natural resources, and for a federal control of every sort of transport and every means of communication throughout the earth. They will claim at length that that is what they always intended – and so fade away...

There must necessarily be a considerable mental aloofness between anyone who is fully aware of the fundamental changes ahead of us and those who are resisting that recognition. He will think well of himself, and it may be he will think too well of himself and too much of himself; but it will be difficult for him to avoid something of the discreet integrity of a man who is fighting a pestilence in a land of dangerously superstitious savages. He may need to get his sanctions from the tribal chief; he may have to allay the jealousy of the medicine men and permit them to claim credit for the miracles of science. Within limits. Tolerance may easily become an enervating cant, a lazy acquiescence in falsification; civility may degenerate into deliberate deception. In his heart, however strongly disposed he may be for superficial compromise, a man sensible of the revolutionary urge of our times is bound to regard a bishop or a contemporary statesman as somewhat of a barbarian and morally and mentally his inferior, and it is difficult to conceal that persuasion altogether. More and more these people must feel the irony that underlies the civility with which they are treated. It is hard to believe that they have been able to get round or over or through the difficulties that are summarized in the concluding footnote to Chapter Three with a really clear and honest mind. It is difficult to talk to them as equals. They make us uncomfortable – for them. Quibbles disregarded and empty churches reproach them for the evaporation of the evangelical fervour of the good old ages of faith. To which, they begin to see more and more clearly, there is no return.

Most of them, as the phrase goes, are "men of the world." They will perform miracles of restatement.

CHAPTER TEN

The Religion of the New Man

The creed of the new religion which is destined to bind a regenerate world together is clear and simple. It demands the subordination of the self, of the aggressive personality, to the common creative task, which is the conquest and animation of the universe by life. The new religion soothes the innate fears and restrains the natural egotism of the young. It repudiates the idea of Sin and the idea of Personal Immortality; which are both in their several ways begotten by Fear. It denies the existence of an anthropomorphic God (Peccavi), and it cannot afford to recognize any prevaricating use of the word "God." That word implies a Personality or it implies nothing.

The new religion maintains that the frame and whole object of life is the progressive conquest of hunger and thirst, of climate, of substance, of mechanical power, of bodily and mental pain, of space and distance, of time, of things that have seemed lost in the past and of things possible in the future. Continually the species will extend itself throughout a constantly ampler and more consciously living universe.

That will go on for ever, since "infinite" is merely a negative term, meaning nothing whatever. It is no good dressing it up in

61

capital letters, calling it The Infinite One or The Absolute and pretending it is something positive and personal. It remains a negative term. The universe, the frame of life in space and time, expands with our knowledge, but it expands without apparent limit.

To that ampler frame the individual life is entirely subordinate, but within the subordination of this religion, and subject to its seriousness and dignity, it will pursue its own course.

The life of the individual will begin at a maximum of illusion and personality; the young will be naturally and necessarily more self-conscious and romantic than the old. In a world of triumphant life the deaths of children and of young men and women will be rare accidents, unless they occur because gallant risks have been taken. But to be killed at the crest of a brave discovery is not so much dying as bringing life to an early climax.

Social morality in the coming order will recognize the phases of life. The young will still have to find themselves in the work they want to do and in their general behaviour. The child begins life protected and acutely aware of its need to possess a protector. The boy is still a hero-worshipper, ready for leadership and reciprocal devotion. He grows up with a God-shaped void in his mind, and it exposes him to much exploitation. At the same time he is acutely aware of himself, so that he may believe in God and yet resent him. He may bargain with God for answers to prayer. He may disregard him, return to him, but it is doubtful if many human beings achieve the disillusioned self-reliance of maturity before their thirties. And also he will desire to be loved and to dominate the woman he desires. Who will reciprocate with her own ardent egotism.

The free, self-centred passions of the young will pass on, tolerated and wittingly, to the maturer preferences of the years of citizenship. The ultimate reabsorption of the individual will not rob the life of the individual of a particle of freshness and interest. The conquest of ignorance by mankind will bring all available knowledge within easy reach of everyone, but that does not mean an oppressive omniscience for the individual; it means merely a delightful progressive discovery of the world and one's own powers. "This I can do better than most people; this I would like to do; this I will do," will be the formula of living.

There is not an individual alive now who could not do something better than most people, if he or she had the opportunity to do it. The conquest of social order will make such opportunity the common lot. The people we tick off as stupid and useless today, born inferiors and so forth, are merely people baffled from the start, people misapplied; the common criminal is generally a rather ingenious person who has lost his way early in life and fallen into social rebellion. The jail reproaches the presumptuous incapacity of the educational system. In a world of intelligent Euphoria – in a world, that is to say, of easily accessible good nourishment for mind and body – there will be little or no need for jail or schooling. Healthy human beings will want to know and want to do. Normal people free from fear will learn by a natural appetite, and go on learning as naturally as they eat and go on eating.

Gradually as the individual life unfolds from youth to age it will achieve satisfaction. The old will become friendly spectators in a sympathetic world; their sense of identification with the species will intensify as they ripen. Life will broaden out from the intensive contemporaneousness of youth, when every trifling event seems of inexhaustible importance, to the broad vision of age in which the whole story of man for the past ten thousand years is yesterday and the next century or so the dawn of a greater tomorrow. We discussed this subjective broadening-out of the Now at the close of our first chapter, and we return to it here. The full course of human life is from action to an ultimate serene and impersonal contemplation. The body and the personality will die at length like an old garment laid aside. This is no dreamland vision. This is the new way of living which is now being born, out of this superficially chaotic conflict. It is a way of living that one can anticipate and share to a very considerable extent even now.

The wild uproar and the stabbing demands and dangers of the present time should not blind us to the fundamental insignificance of these events. These are the birth-pangs of the human release. "Sorrow and sighing may endure for a night, but joy cometh in the morning." The stars in their courses fight for the new humanity. The reality of human history flows on beneath the troubled surface of these accidents. These conflicts may seize upon our individual selves and oblige us to risk or lose our personal time, our personal work, and our personal lives; but that must not blind us to the incidentalness of these occasions that have entangled us. War and civil conflict draw more and more of us into the vortex of killing, and in many instances there seems to be no alternative to killing. Non-resistance is a treachery to all mankind and not simply to ourselves. Anyone may have to defend himself and others at any time from a homicidal lunatic. You may have to kill and risk being killed.

But we must not mistake even a vast plenitude of such individual distractions for a world disaster. These governments and boundaries, the religions, hallucinations, and antagonisms of the still unenlightened past, the British Empire, the Catholic Church, the American Way of Life, La France, the Germanic Race, the fooleries of political "hegemony," and so forth and so on, encumber our world today; we have still to live for a time among these guttering ruins; too often we have to judge and take sides with a lesser evil against a greater; but so long as we do not treat these warring "Powers" and "Faiths" as primary and permanent realities, so long as we despise them utterly in our hearts, although we must use the best of them as stepping-stones to reorientation, we can still keep our contact with the new way of living and know the world is ours. These discolourations of the human mind will fade as the light grows stronger, and ultimately they will fade out of human consciousness altogether. And as our species conquers time it will reach back to realize more and more exactly, and to live again more and more fully, the contribution it has incorporated from our lives.

Appendix I

Footnote to Chapter One

The Increase in Human Facilities

I n *World Brain* (Methuen and Co., 1938) and in *Science and the World Mind*, my 1941 British Association Address, which is now reprinted by the New Europe Publishing Company, 29 Great James Street, WC1, the reader will find some of these current and possible extensions of human facilities amplified. At the present time it is particularly difficult for publication to keep pace with the progress of invention. Anyone anxious for a fuller documentation of what I have written about photomicrography may be referred to the bibliography in the handbook issued by the Eastman Kodak Company of Rochester, New York (*Photomicrography: An Introduction to Photography with the Microscope*), containing thirty-four titles and bringing the literature up to 1935. But the object of the handbook is obviously to inform purchasers of material the company has available for their use, and when I discussed the possibilities of the camera for record and distribution with my old friend Dr Kenneth Mees (we discussed *Anticipations* forty years ago), his suggestions of what could be done ran far ahead of any published intimations.

In all these matters there is serious conflict between business considerations and progress. New inventions are held back from common knowledge until unsold stocks can be disposed of and plant reconditioned. When I ask "His Master's Voice" people to tell me what we may expect from them when skilled labour takes up the gramophone again, I am told with great dignity that the time is not ripe for any such intimations. There is reason to believe that a new type of record and apparatus is now possible which will give a perfect rendering of an opera upon one disc, just as it seems possible that already any scene caught by a motion camera may be televised with a minimum of delay all round the earth. But most of us will never be permitted to see these things arrive in our lifetime.

In a different social order invention and discovery would not be locked up and held back in the carefully guarded laboratories of great profit-making concerns; the research work of the modernized university laboratories would be endowed far beyond the resources of the contemporary business world. But all these devices, organizations, and enhancements of human capacity which would be eagerly welcomed and anticipated in a mentally active community are deprecated and depreciated with the utmost earnestness by our mandarin-educated rulers. Their sense of superiority, their cultivated defensive conceit, is invincible. They do not seem to care even for the happiness of their own children. They can tolerate no other order of thing than the one which gave them their own importance. Their idea-resistance is so powerful that it will probably delay the escape of the masses of mankind from fear, privation, and needless servitude for some unhappy generations.

Confronted by the gathering evidence that it may be possible to produce practically unlimited quantities of foodstuffs by the exposure of nutritive solutions to intensive illumination, that the ploughman may cease to plod his weary way homeward and the herdsman and shepherd turn to other occupations, while a vast variety of old-fashioned and new-fashioned foodstuffs will be abundantly available everywhere and all the year round, they give way to storms of angry detraction. When the chemist tells them that it may be unnecessary to pass grass through the body of a cow before it is fit for human consumption, their distress is uncontrollable. Nevertheless "the lowing herd winds slowly o'er the lea" and exit left; and, whether they like it or not, "leaves the world to" a working day of twenty-four hours in four or five hour shifts under artificial illumination of the most favourable sort and – the pensive observer – "me."

In this matter of the world mind, for example, I have before me Paper No. 163 of an American biological publication *Growth* (Vol. 5, No. 2, 1941). It contains a list of thirty-nine other references. In it Zamenhof deals with the possibility of greatly increasing the functional efficiency of the brain cortex by feeding the developing individual with particular growth hormones. The nerve elements and nerve connections under that treatment become finer and more abundant. In other words, a new sort of brain with the power of carrying on a far more subtle thought-process, over and above the external economies at which I have glanced in *World Brain*, is within the range of human possibility. You would imagine that the whole world of teachers, psychologists, and so forth would be agog at these tremendous intimations of human release. Nothing of the sort will occur. They will resent them as a personal affront. Discreet young men with an eye to promotion will practise evasions of such exciting questions until it becomes a second nature.

Long years ago I had a conversation with Sir Michael Foster about the possibility of delaying the closing-up of the sutures of the human skull and so permitting the grey matter of the brain a longer period of expansion, with a consequent prolongation of mental development. "You can write about this sort of thing," he said, "because you need not seem to believe in it, but I dare not say a word about it if I am to remain Member of Parliament for the University of London." And so our learned mandarins do their best to emulate the brontosaurus and stick themselves (and us) in the mud until the stars in their courses turn against (us and) them.

APPENDIX II

Footnote to Chapter Eight

A Summary of Modern Ideas about Space and Time

Four-dimensional geometry has been an interesting exercise in speculative mathematics for a century or more, and became exciting and popularized by the Relativity boom some twenty-odd years ago. Four-dimensional geometry was a subject for discussion among my fellow-students at South Kensington half a century ago. One or two of us concluded definitely that this fourth dimension of the mathematicians was duration. We argued in this fashion: "Nothing material exists instantaneously. It must have length, breadth, thickness and – duration. The fourth dimension is duration."

I don't remember that we followed up the consequences of that conclusion with any severity. Later on I wrote a fantasy called *The Time Machine*, in which a machine travels *through* time. It was entirely fantasy, and the reader was bluffed past the essential difficulties of the proposition entirely for the sake of the story...

I met J W Dunne just before he went off to the Boer War because he had been making some experiments about flying and he wanted me to take care of his results if anything happened to him. Nothing did. He returned and built one of the earliest machines that ever went up into the air. But presently he got himself badly infected with the four-dimensional idea, and he began to think and write about it. I do not think he thought very well in this particular field. The storyteller's bluff had deceived him. He accepted its suggestions as a man in the audience takes a forced card from a conjuror.

Coming down to hard statement, the objection I have always made to Dunne's line of argument is this. In a two-dimensional world you have a north and south axis and an east and west one. But between them you have an angle of 90 degrees. You can speak of NE, NNE, N by E, NE by E and so on. Go on now to three dimensions, and your degrees and latitude give you a similar number of intermediate degrees. From the centre of a sphere you can follow the rotation of a radius to any point of its surface, and by the formulae of spherical trigonometry you can state its position precisely. Your ship, for example, can be located exactly as being so many degrees west of Greenwich in latitude 40° N relatively to the centre of the globe.

But now pursue this analogy or identity of time with these other spatial dimensions. The time dimension, according to our assumption, should be *past and future* axis at right angles to the other three. In which case there must be intermediate angles between it and the other three axes. Where are they? *Apparently they have vanished.* If you will read Dunne's book it is evident that he failed to observe that. But if they are not there, then time is *not* a dimension like the three spatial dimensions, and that serial universe of his is a delusion based on a false analogy.

Now, clinging to the conclusion of my student days that nothing can exist without length, breadth, thickness, and duration, I found myself groping about to discover what had made those intermediate angles practically imperceptible – because I still believed they were there. And presently I perceived that this linked up with another remarkable thing about our universe, and that was that there was an absolute limit to the speed at which things could move in it. That limit was the speed of light. The Michelson and Morley experiments, with their repetitions and confirmations, established this limitation as incontrovertible.

In our universe the extreme limit of motion is 299,796 kilometres a second. Nothing has even been known to go faster than that. You can imagine yourself doing so, you can imagine yourself on a rocket going at 299,796 kilometres a second, and you can imagine yourself standing up and leaping forward so as to go faster than that alleged maximum. But in reality you cannot do that. You are imagining the impossible.

Here we come upon a matter in which human beings vary very widely – their power to revise their fundamental ideas in the face of new facts or novel considerations that invalidate those fundamental ideas. There are a great number of people who seem totally unable to readjust these primary assumptions; there are others who resist, consider the new facts, and then carefully and completely readjust; there are a few who seem to grasp the new way of looking at things so readily that they adapt without a struggle.

No doubt there is an inherent quick-wittedness in the latter sort, but it is also possible that the former kind are in many cases the results of educational repression and an acquired inertness of the mind. Space and time, as they knew space and time, had become the *nature of things* for them, and it was inconceivable to them that anyone should argue about them. But those others knew that their convictions about space and time were merely interpretations of the evidence of their senses, and that a different interpretation, or many different interpretations, might be possible.

We may cite various instances in which the mind revises the sensory impressions it receives. You see a man walking right way up. But in fact the image on your retina is wrong way up, and it is your mind that supplies the proper interpretation. Your baby learns the interpretation of its senses by a thousand experiments which it forgets as soon as it has acquired a working code. At first it squints, has no sense of distance, tries to take hold of the moon. Until we set about teaching it things – we call it teaching – it lives in a limited flat world of familiar things and extends its knowledge by analogy.

Then it begins to be taught. It learns that there is a "home for little children above the bright blue sky" – it is pointed *upward* to Heaven; it is told the sun "rises in the east and sets in the west," and simultaneously that "the world is round like an orange."

These incongruous ideas are never brought into conflict, and it is soon taught by example that it is better to get exasperated by disturbing questions and drive them away rather than ask or answer them. So its ideas of a Heaven overhead and of a world that is round like an orange get accustomed to lie side by side without quarrelling. It never develops a critical habit of mind. It finds that questioning things is a burthensome fatiguing process that hinders rather than helps in the ordinary affairs of life and may easily provoke social dissensions.

The young citizen grows up with a sort of defensive muddle in its head and a defensive habit of mind that works fairly well in a muddle-headed society not in immediate danger of disaster.

Yet even in a world that is a going system of established things a certain number of minds are more or less troubled by that suspicion that there may be a catch in it somewhere, that the whole display may be capable of a different interpretation. This uneasiness is often personified as the Demon Doubt. It hits at Revelation and the omniscience of priests and teachers. Those in authority hate it and do everything they can to exorcise it. But in a certain proportion of minds curiosity still lives, and the genuine scientific investigator, regardless of morality and public policy, would rather die than leave that queer little inconsistency alone. The history of human liberation is largely a progressive eating-away of confident common-sense assurances about existence, by little gnawing persistent facts. That, for example, is the history of astronomy. Little facts that every commonplace person *knew* were nonsense gradually dislodged the planetary disc from its position at the centre of the universe, rolled it up into a ball, sent it spinning smaller and smaller through the expanding galaxies. To the great irritation of commonplace people...

The story of Relativity[1] has only repeated that time-honoured theme. These four-dimensional mathematicians had been playing with four-dimensional geometry for the better part of a century, and they were, I believe, very indifferent to the revolutionary possibilities of the stuff they were handling. The world at large had the Newtonian framework made up of three spatial dimensions *and* time *and* gravitation, and most educated people in the world were perfectly content with that framework. That was the *nature of things* to them. They could not imagine any alternative to it, and it seemed to matter to nobody except a few specialists that speculative mathematicians were playing with the complicated possibilities of a geometry of our spatial dimensions. In moments of condescension these latter would make their ideas quite attractive and quite harmless-looking to ordinary minds.

You could begin, they pointed out, with a geometry of two dimensions, a Flat World. [2] People, flat people, living in a flat world, would be quite unaware of thickness. But if someone living in three dimensions were able to lift a flat individual out of his flat world and then put him back in it, it would seem to other flat people in that particular world as though he had vanished and returned. Moreover, if the three-dimensional person turned the flat man round before putting him back, his right hand would now be his left and generally he would be inverted. And it will be plain I think that all possible Flat Worlds, an infinity of them, could be put side by side like the pages of a book in a three-dimensional universe.

It is necessary to be very clear about the determination of position in any sort of universe. You have a dot, and in a one-dimensional universe you locate it by saying it is so far ahead or behind, and that is all that you can do in that linear system. But now take a two-dimensional universe. You have now, ahead and behind, which is one axis, and right and left, an axis at right angles to that. You are a flat individual and you want to define the position of a dot in your flat world. It is, we will suppose, ahead of you and to the right of you. You define it by referring it to your fore and aft axis and your right and left axis. It is so many degrees ahead on the right. If it moves you can make a trace of its movement in reference to right lines – to linear coordinates, that is to say, coinciding with these axes. Every point and every movement in your entire universe, excluding nothing whatever, can be indicated in this way, and cannot be indicated so simply in any other way. No other way is necessary, and it is the scientific method to exclude every superfluous hypothesis.

But I have heard it suggested: "Cannot I turn these axes round and measure from other coordinates?" You cannot, because you will have to turn them to your left or your right, north-west or north-east. North will still be ahead of you and west on your right hand. But you say: "I don't want *that* universe; I want a different one; I want one with *your* north on my right." Then yours will be a different universe relative to mine. The traces made by moving objects will be different. You will have to stick to *your* dots and traces, and it will be a complex problem to calculate their relation to mine. You will have done yourself no good at all, and you will be just as bound to *your* universe as I am to mine. If you are freakish you may want to spin into other universes. Each, you will find, will be as inexorably a complete and separate universe.

And similarly, if you are a three-dimensional being, the three axes of Newtonian space hold you bound. The rotation of the earth gives a most convenient system of three axes in Newtonian space: north and south, east and west, and up and down. There is nothing to forbid your making Sirius your pole star if you like, but that will merely bring you to the verge of brain fever if you want to define position and movement. One of the axes of your three dimensions is not strictly at right angles to the others because it is radial from a centre, since Copernicus and others have curved the original levelness of the earth's surface. But there it is, a completely definable universe, to which all similar universes are relative. It is not an unlimited plane, but it is still an unlimited sphere.

In this fashion these mathematicians led us along the path to a four-dimensional world. Because, said they, if there is also a fourth dimension, which there may or may not be, then similarly you could put an infinitude of three-dimensional systems side by side in a four-dimensional universe. And they proceeded to work out the intricate geometrical problems this gave them, the problems of endless Newtonian worlds side by side. I do not think they worried very much about whether the fourth dimension was time or not.

I exploited this idea of endless Newtonian worlds side by side in a story called "The Plattner Story" (1896) and also in a book, *Men Like Gods*, where a carful of people skidded, by means of a little literary hokey-pokey about a "kink in space," into an adjacent world lying like another page in a book, side by side with ours. (Men had done better in that exemplary world, but they wore no clothes, and that was a great bother for the illustrator of *Hearst's International*, whose editor had bought it rather carelessly for serialization. But that is by the way.)

You can express selected aspects of our four-dimensional life in one, two, or three dimensions. By confining yourself to the time dimension you can express life as music and all the possibilities of language, by using two dimensions you have pictures and inscriptions, with three you can have either the solid model, statuary, architecture, or else by bringing in time and omitting one of the others you can have the talking movie, and by resorting to all four you can express yourself in drama, opera, ceremony, parade, dance or demonstration. But that is the limit set to us four-dimensional beings. Expression in five dimensions, existence in five dimensions, is totally inconceivable.

That brief explanation will I think make matters at least a little easier for the reader unaccustomed to this sort of speculation. The Two-Dimensional universe makes a universe without a third dimension conceivable, and that again by analogy makes a four-dimensional system thinkable.

But now we introduce a fresh aspect of this question. So far in this explanation I have dealt only with the idea of a four-dimensional geometry in which there are countless three-dimensional universes like the leaves of a book. But now let us take up the suggestion that the fourth spatial dimension is time. It is a spatial dimension but with rather different relations to the other three. That still leaves it conceivable that there are innumerable three-dimensional universes lying side by side, but it introduces the idea that our particular universe, our "nature of things," is a four-dimensional one, and that we are four-dimensional creatures. We are, it is suggested here, four-dimensional beings, no more and no less. There may also be a plurality of four-dimensional universes and so on, but that is another story for which we have no mental equipment. It is not the story of the universe in which we exist.

What follows if we recognize that we are living in a four-dimensional universe? That question revives the difficulty which I have already broached in my opening remarks about Dunne. I return to that. Where, I repeat, are those missing angles, North by Past, North by Future, Future by West, Future by the Tropic of Capricorn? Why are they practically imperceptible?

An explanation has to be found if we are to justify that assumption. It is possible to make quite a plausible one, and that we will now do. But bear in mind we are dealing only with plausibility and not with absolute veracity, which seems to be quite inaccessible to minds such as ours.

Suppose that we represent our visible *universe* as a three-dimensional system – that is to say, as a solid universe that would become like an immovable waxwork model if there was no such thing as time. And suppose that instead of being so motionless, as we represent it in paintings and photographic stills, it is really flying through the space-time continuum along the time axis at this terrific speed of 299,796 kilometres a chronometer second, then the shiftings-about of ordinary events (of the objects, that is to say, in that quasi-stagnant three-dimensional spectacle) will be so slow in comparison with its movement as a whole along the time axis as to make an apparent difference in kind, a quantitative difference which is apparent rather than real, between the normal events of everyday life and the universal event of the passing of time.

Our flight along that past and future axis is as if we were travelling at that incredible pace on some Coney Island Speed Railway or sitting in some equally superlative dive bomber. We cannot get off and stretch our legs and pick flowers and get in again. That sort of speed seems like one thing and walking along a lane another. But they are really both speed. And this analogy may perhaps be bettered by saying the whole of our visible universe is boxed in, and we are all together for this flight we are making along time at the rate of 299,796 kilometres a second. We move about in this headlong vehicle as a passenger can walk about in an express train, but his movements in the line of flight are relatively so inconsiderable as to be imperceptible. Most of his movements will be lateral – more or less sideways, that is, to the flight of time. By jumping down a precipice he may attain a velocity of a few metres a second while he will have fallen 299,796,000 metres from past to future. The muzzle velocity of the biggest of guns in the cleanest of states is hardly more important in relation to the flight of time.

This enables us to conceive of ourselves as four-dimensional beings in a four-dimensional universe. Nothing exists in our universe as it is presented here without length, breadth, thickness, and duration, and this conception that time is purely spatial can give us as satisfactory a frame for the *nature of things* as man is likely to attain for some time.

The telescope ultimately obliged men to realize that the earth was not flat, and that in place of accepting three spatial dimensions absolutely at right angles to each other they had to recognize that there was a curve in the system. They learnt that by going east on the earth they came round again to west, and that the truer picture of the *nature of things* was a picture of great circles returning into themselves. The requirements of a fourth-dimensional spatial system now demand a still more complicated frame of coordinates (the Gaussian coordinates). This the reader will find discussed with the utmost lucidity in Einstein's *Relativity*. To embark upon a condensation, if a condensation of that already very compact crystal of reasoning were possible, would carry us outside the scope of this work, in which the nature of physical events is not a primary issue. Here it is necessary only to say that the picture of the physical nature of things Einstein develops is of a universe that completely returns into itself, and is therefore *finite*. It is four-dimensional; but if we want a crude three-dimensional parallel of it, it is a rigid system of unique events shaped rather like a dumpling. It is the whole of existence. It is the all.

This finite four-dimensional universe in which we live and move and have our being can expand. But if you find yourself believing it is expanding into some pre-existing space that was previously empty, then you have failed to grasp the four-dimensional idea. You are still under the spell of what Einstein calls the "Galilei–Newton" conception of three-dimensional infinite space. There is nothing whatever outside our four-dimensional universe, neither space nor time. *All* space and time are in it. That universe expands by its material events getting farther apart. If you want to get this clearer, you must be referred back to your studies; you must go back to Einstein's book and to Eddington's *The Expanding Universe*.

It may be well for greater clearness and emphasis to repeat what has been said here. The past and future *exist permanently* in this universe, and our consciousness is a series of delusively unified conditioned reflexes. That illusion of a unified personality joins up a series of traces which constitute our conscious life, and this conscious life *takes the form* of a fall or flight, along the spatial dimension we call time. It is a rigid form. From the standpoint of the space-time continuum there is no movement; the whole system is rigid. It is simply from the subjective and illusory point of view that there seems to be free movement. The four-dimensional universe is rigid, Calvinistic, predestinate; the personal life is not a freedom, though it seems to us to be a freedom; it is a small subjective pattern of freedom in an unchanging all. There is no conflict between fate and free will; they are major and minor aspects of existence. The major aspect of life is Destiny; the minor is that we do not know our destiny. We struggle because we must; and that struggle *is* life; but the parts of the drama we enact belong to a system that has neither beginning nor end.

This Appendix will make the profoundly revolutionary nature of dimensional discussion understandable for those who are unfamiliar with these questions. But many minds, either by bad training or natural ineptitude, are absolutely incapable of dimensional criticism. They cannot disentangle themselves from their ingrained ideas, and since they cannot cast out or suspend the old, they are incapable of taking in the new. They try to think about these matters in the forms of thought that they repudiate. They do not realize that this discussion deals with what they call material things, and they slide off happily into some oracular wisecrack that the Fourth Dimension is God or the World of Spirit or the Day of Judgment or anything rather than the past and future axis in the time-space continuum.

Others again grasp these ideas but do not grasp them very firmly. Very readily they will slip back again to some long-established preconception in their minds, that there is, for instance, an antagonism of matter and spirit and that this modern religion is not spiritual. Most emphatically it is not. Nor is it material. There are no ghosts of any kind in this book. It is monistic.

I hope this will clear up some of these notions for the general reader, but it may be well to return to quite another point that has been dealt with perhaps too allusively in the main text. There is the question of the "Now" which we raised at the opening. Ray Lankester considered that the "Now" was when the heartbeat carried the freshly aerated blood into the brain and stimulated it to consciousness. For that moment, for that "Now," the mind exists. Then it ceases to exist until the next heartbeat.

Our conscious lives are like the pictures on a cinema screen. They are discontinuous "Nows," but they follow one another so rapidly that they seem continuous. That is our waking life. With the ebb of the stimulating oxygen, the presentations sink below the level of consciousness. Our conscious life ceases. Some pain, some noise, morning, brings back the vivifying flow, first of all as a confusion, without any rational sequence, which our awakening mind seeks to rationalize as it fades. The return to reality is usually discontinuous with the last conscious phase. It picks up some earlier memory.

Often after some violent shock there is an effacement of the memory of the event leading up to the shock. You find yourself in the hospital. You are told you walked in front of a cab and were knocked down, but the last thing you recall is a shop where you were making purchases. Minor lapses in continuity of the same essential nature are of frequent occurrence, but, since we do not recall them, usually we do not know of them. It is not that the events are not seen or heard, but that they do not join on. They may crop up perplexingly in dream-stuff or mingle with scenes we have imagined and rehearsed.

So we are pieced together, and so we piece our lives together as our hearts beat. As we have already shown in Chapter Two, our personalities are merely "serviceable synthetic illusions of continuity." And, at the last throb, a soft dark restful curtain falls for ever upon that personal life and our contribution has been made.

1 Of which quite the most lucid account is to be found in Einstein's little book entitled *Relativity*.
2 Or with a geometry of one dimension, in which your universe would be a line with only backwards and forwards in it.

The Happy Turning

CHAPTER ONE

How I Came to the Happy Turning

I am dreaming far more than I did before this chaotic war invaded my waking hours. My days are now wholly full of war effort: What can I do? What ought I to do? Where is the next opportunity and what dangers gather ahead? I am urgent. I overstrain. And now something deep within me protests and rebels, and says: "These war-makers have yoked and enslaved you. You are defeated if you give yourself wholly to war."

I answer evasively: "Presently I will relax."

That serves in the daytime but not at night. I take care to keep as fit as I can and not to let my war preoccupations develop into the nervous waste of anxiety. I never dream about the war. I dream neither of its horrors nor its strategy. When I sleep, a more adult and modern and civilised part of my being comes into play. More and more are my dreams what I believe the psychologists call compensatory; the imaginations I have suppressed revolt and take control.

Some time ago I dreamt a dream that recurs with variations again and again, so that it is a sort of Open Sesame for all my excursions into dreamland. In my daytime efforts to keep myself fit and active, I oblige myself to walk a mile or so on all days that are not impossibly harsh. I walk to the right to the Zoo, or I walk across to Queen Mary's Rose Garden or down by several routes to my Savile Club, or I bait my walk with Smith's bookshop at Baker Street. I have to sit down a bit every now and then, and that limits my range. I've played these ambulatory variations now for two years and a half, for I am too busy to go out of town, out of reach of my books, and my waking self has never uttered a protest. But now the – what do they call it? – subliminal consciousness? – has in the most charming way asserted my unformulated desire, with this dream, which I will call the dream of the Happy Turning.

I dream I am at my front door starting out for the accustomed round. I go out and suddenly realise there is a possible turning I have overlooked. Odd I have never taken it, but there it is! And in a trice I am walking more briskly than I ever walked before, up hill and down dale, in scenes of happiness such as I have never hoped to see again. At first the Turning itself was the essence of the dream. Now directly my dream unfolds I know where I am; it has become a mere key to this delightful land of my lifelong suppressions, in which my desires and unsatisfied fancies, hopes, memories and imaginations have accumulated inexhaustible treasure.

For the first time in my existence I realise what it is to have possession of an entirely healthy and balanced body. I was born astigmatic and in those days nobody bothered about common children's eyes. I could never be sure of bowling a straight ball, and when I jumped down I hit the ground too soon or too late. I was under-nourished and tuberculous, so that I was a skinny puny youth, easily fatigued. Tolerable health came only in my thirties. Muscular precision and hardiness I shall never know in my waking life. But now, beyond the Happy Turning, I leap gulfs unerringly, scale precipices, shin up trees and am indefatigable. There are no infections there; no coughs, no colds; to cough or sneeze would be to wake up and tumble back headlong into those unhygienic present-day realities where dirt-begotten epidemics have their way with us. Maybe a day will come when a cleansed and liberated world will take the Happy Turning in good earnest and pass out of the base and angry conflicts which distract us from wholesome living. All such liberations are possible beyond the Turn. Now I count it good fortune that I can even dream of the gay serenity of that Beyond.

The Happy Turning leads to a world where distance is abolished. Certain phrases – parroted phrases empty of belief – are already to be found in the newspapers and speeches – the abolition of war, the abolition of distance, the abolition of competition and social inequality. But after people have repeated a phrase a great number of times, they begin to realise it has meaning and may even be true. And then it comes true. Beyond the Happy Turning these phrases are realities; hopes fulfilled.

CHAPTER TWO

Suppressions and Symbolism in Dreamland

But the fantasies of dreamland go an immeasurable way beyond what is now conceivable and practical.

The subliminal self is never straightforward. It awakens, us, for example, to sex and the social reactions of adolescence in the queerest, most roundabout way. There are sound biological explanations why our minds should work in this fashion, but I cannot go into them now. The submerged intervener is cryptic and oracular; it hints and perplexes. Symbols become persons and persons symbols; individuals, animals, institutions, amalgamate and divide and change into one another.

Religions are such stuff as dreams are made of. The Athanasian Creed is severely logical in dreamland, Isis is transfigured into Hathor, a cow, Quannon, the crescent moon and Murillo's Queen of Heaven, and still the dream flows on. Osiris becomes his own son Horus, who becomes again Osiris and the Virgin Mother, in incessant rotation. This is the atmosphere of this uncontrollable Wonderland beyond the Turn, in which my accumulated loves and suppressions, disappointments and stresses, find release. But very plainly it is my personal needs that provide the substance of the stories with which my dreaming self now consoles and regales me.

In the past I do not recall dreams as a frequent factor in my existence, though some affected me very importantly. As a child I used to have a sort of geometrical nightmare as if a mad kaleidoscope charged down upon me, and this was accompanied by intense distress. I may have been very young then, because I cannot remember how I awakened or whether I conveyed my distress to anyone. Nor have I ever come upon a description of that dream as happening to any other child.

But I remember a considerable number of quite frightful dreams that came before my teens. I read precociously, and I was pursued implacable, to a screaming and weeping awakening, by the more alarming animals I read about. An uncle from the West Indies described some frightful spiders that scratched and crawled. I was then put to bed alone in the dark in the upstairs bedroom of a strange house, and I disgraced myself by screaming the house down.

I had horror dreams of torture and cruelty. One made me an atheist. My mother was a deeply religious woman, but she did her best to conceal the Devil from me; there were pictures in an old prayer-book showing hell well alight, but she obliterated these with stamp paper which I was only partially successful in removing, so that until I held the page up to the light, hell was a mere suspicion. And one day I read a description in an old number of *Chambers' Journal* of a man being broken on the wheel over a slow fire, and in my sleep it flared up into immeasurable disgust. By a mental leap which cut out all intermediaries, the dream artist made it clear that if indeed there was an all powerful God, then it was he and he alone who stood there conducting this torture. I woke and stared at the empty darkness. There was no alternative but madness, and sanity prevailed. God had gone out of my life. He was impossible.

From that time on, I began to invent and talk blasphemy. I do not like filth. Merely dirty stories disgust me, and when sexual jokes have an element of laughter in them almost always it is dishonouring and cruel laughter. But theology has always seemed to me an arena for clean fun that should do no harm to any properly constituted person. Blasphemy may frighten unemancipated minds, but it is unbecoming that human beings should be governed by fear. From first to last I have invented a considerable amount of excellent blasphemy. *All aboard for Ararat* is the last of a long series of drawings and writings, many of which have never seen and probably never will see the light of print. There must be lingering bits of belief in order to produce the relief of laughter, and such jests may fade out very rapidly at no very distant date.

Only a few other dreams stuck in my memory before I discovered the Happy Turning, and mostly they were absurd and misleading freaks of fantasy. I dreamt my mother was ill and in great distress and wrote off post haste to her. There was nothing at all the matter with her.

I must have had anxiety dreams when I was over-working, in which everything was at sixes and sevens, I must have had them because I devised a technique for dealing with them. Directly I woke up, I got up and dismissed them. I trained myself to make tea and set to work soberly in a dressing-gown, and soon everything fell back into its place and the disturbance succumbed to fatigue and natural sleepiness. My friend J W Dunne, who wrote *An Experiment with Time*, lost himself for a time in a Serial

95

Universe and has come back a most delightful writer of fantastic tales, induced me to keep a notebook at my bed-head and jot down my dreams fresh and hot. I do not remember making a note. I just woke up, and whatever dreams may have been hanging about vanished unimportantly forthwith.

So my present resort to dreamland is a new experience. I am a happy explorer telling of a delightful world he has come upon, beyond expectation.

CHAPTER THREE

Compensation Beyond the Happy Turning

The scenery of my dreamland is always magnificent or exquisite or otherwise delightful. I should not note it if it were not, and I find dear and delightful people I had never hoped to see again, happy and welcoming. Sometimes they are just themselves for a time, sometimes they are agreeably blended with other people, and at any moment they may see fit to impersonate someone else and cease to be whatever they began by being.

Nobody is dead in this world of release, and I hate nobody. I think that this absence of hate may be very recent. It may be due to my subconscious revolt against the unavoidable hates, disputes, suspicions and conflicts of our daily life in this war. Or it may be that with advancing years a mellowing comes to the mind with attenuation of ambitions and rivalries. They matter so little at seventy-seven. Both factors, the normal one and this abnormal one of war conditions, may be contributing to my escape.

My waking life is now one of very fierce and definite antagonisms. I feel that the generations ahead may be cheated of much or all of the huge emancipations that could and should follow upon this world storm of fighting; and that ancient and evil organisations and traditions and the necessity common minds are under to believe they have natural inferiors, of whom they are entitled to take advantage, may frustrate all our hopes. I am compelled to spend my utmost energy in warfare against these things.

Dreamland is in flat contradiction to all this distressful strain. Nothing of these conflicts pursues me beyond the Happy Turning. At the Happy Turning is a recognisable Holy Water Stoup which has somehow identified itself with Truth, and in my Dreamland there is not the slightest difficulty about dipping a finger and sprinkling the Holy Catholic Church, or whatever ugly menace to mankind happens to be upon my heels, with it. Whereupon the evil I fear and fight here with all my strength, explodes with a slightly unpleasant odour, and vanishes. Why did I let my heart be troubled? Why was I afraid?

CHAPTER FOUR

The Holy Carnival

Nothing distressful to me can clamber over that Threshold now. But anything and everything that shows me deference may play its part in my relaxation. I have had some very entertaining divine conferences. The gods men worship are difficult to assemble and impossible to count, because of their incorrigible habit of dissolving spasmodically into one another. I have remarked already upon the permutations and combinations, if those words are permissible, of Isis, the original Virgin Mary. Cleopatra's infinite variety was nothing to it. The tangle of the Trinities is even more fantastically versatile. There is the Athanasian Trinity and Arian Trinity, the Catholic and Orthodox, the Logos and that ever ambiguous Virgin. There is the Gnostic Godhead, which makes Jehovah out to be the very Devil, and Pope's consolidated Deity:

"Father of all, in every age, in every clime adored,
By Saint, by Savage and by Sage, Jehovah, Jove or Lord."

The vast theogony galumphs about in an endless confusion of identities with a stern transcendent solemnity that never deserts it. "Which except a man believe faithfully, he cannot be saved."

A few such cries are uttered with an air of profound significance; a considerable amount of thunder goes on, a crackle of miracles, but never a laugh. To laugh is to awaken.

And in and out and round about this preposterous dance of the divinities, circulates an innumerable swarm of priests and prophets and teachers, wearing the oddest of robes and garments, mitres, triple crowns, scarlet hats, coquettish hoods. No Carnival gone mad can compare with this insane leaping and tumbling procession. They pour endlessly through the streets of my dreamland; striking strange symbolic attitudes, some with virgin beards, some grotesquely shaven and shorn, hunchbacked with copes, bellowing strange chants, uttering dark sayings – but always incredibly solemn. They tuck up their petticoats, these grave elderly gentlemen, and one, two, three, leap gulfs of logic.

I noted the present Primate, chief now of the English order of primates, his lawn sleeves like the plump wings of a theological Strassburg goose, as, bathed in the natural exudations of a strenuous faith, he pranced by me, with the Vatican a-kicking up ahind and afore, and a yellow Jap a-kicking up ahind old Pope. I had a momentary glimpse of the gloomy Dean, in ecstatic union with the Deity, yet contraceptive as ever, and then, before I could satisfy a natural curiosity, a tapping delirium of shrilling cymbals swept him away, "Glory!... *Glory!*... ALLELUIA!..."

As, on the verge of awakening, I watch this teeming disorder of the human brain, which is always the same and increasingly various, I listen for one simple laugh, I look for one single derisive smile. Always I encounter faces of stupid earnestness.

They are positively not putting it on, unless earnest self-deception had become second nature. They are not pretending to be such fools. They *are* such fools...

There is this phase between dreaming and awakening, there is a sense of rapidly intensifying conflict and strain before the straining catgut snaps – exactly as it snaps when we come out of anaesthesia. The Brocken Witches' Sabbath begins dispersing and dissolving, becomes a wildly spinning whirl. Will there be enough broomsticks for everybody? Hi broomstick! Are you engaged, broomstick? That's *my* broomstick. They all leap for the nearest one. They rush to and fro about me and through me, terrified at the Berlioz clangor that heralds the night of the Gods. The Archbishop, Inge, His Holiness, Rabbis, thrust about me. They rise in the air like a whirlwind of dead leaves. They spin up towards the zenith colliding and fighting among themselves – serious to the end.

Cosmo Gordon Lang, I remark, gets into theological difficulties with his steed, which rears and throws him. There is a wild struggle in which his broomstick vanishes. Down he goes, legs and arms and robes, cartwheeling faster and faster. The dream becomes a religious hailstorm. Whiz, whiz, they come pelting.

I have a vague idea I ought to put up an umbrella. Umbrella? I laugh and am awake.

CHAPTER FIVE

Jesus of Nazareth Discusses His Failure

The companion I find most congenial in the Beyond is Jesus of Nazareth. Like everything in Dreamland he fluctuates, but beyond the Happy Turning his personality is at least as distinct as my own. His scorn and contempt for Christianity go beyond my extremest vocabulary. He was, I believe, the putative son of a certain carpenter, Joseph, but Josephus says his actual father was a Roman soldier named Pantherus. If so, Jesus did not know it.

He began his career as a good illiterate patriotic Jew in indignant revolt against the Roman rule and the Quisling priests who cringed to it. He took up his self-appointed mission under the influence of John the Baptist, who was making trouble for both the Tetrarch in Galilee and the Roman Procurator in Jerusalem. John was an uncompromising Puritan, and the first thing his disciples had to do, was to get soundly baptized in Jordan. Then he seemed to run out of ideas. After their first encounter John and Jesus went their different ways. There was little discipleship in Jesus.

He played an inconspicuous role in the Salome affair, and he assures me he never baptized anybody. But he was brooding on the Jewish situation, which he felt needed more than moral denunciation and water. He decided to get together a band of followers and march on Jerusalem. Where, as the Gospel witnesses tell very convincingly, with such contradictions as are natural to men writing about it all many years later, the sacred Jewish priests did their best to obliterate him. He learnt much as he went on. He seems to have said some good things and had others imputed to him. He became a sort of Essene Joe Miller. He learnt and changed as he went on.

Gods! how he hated priests, and how he hates them now! And Paul! "Fathering all this nonsense about being 'The Christ' on *me* of all people! Christian! *He* started that at Antioch. I never had the chance of a straight talk to him. I wish I could come upon him some time. But he never seems to be here... There are a few things I could say to him," said Jesus reflectively, and added, "Plain things..."

I regretted Paul's absence.

"One must draw the line somewhere," I said. "In this happy place, Paul's in the discard."

"Yes," reflected Jesus, dismissing Paul; "there were such a lot of things I didn't know, and such a lot of snares for the feet of a man who feels more strongly than he understands. I see so plainly now how incompetently I set about it."

He surveyed his shapely feet cooling in the refreshing greensward of Happyland. The stigmata were in evidence, but not obtrusively so. They were not eyesores. They have since been disgustingly irritated and made much of by the sedulous uncleanness of the saints.

"*Never* have disciples," said Jesus of Nazareth. "It was my greatest mistake. I imitated the tradition of having such divisional commanders to marshal the rabble I led to Jerusalem. It has been the common mistake of all world-menders, and I fell into it in my turn as a matter of course. I had no idea what a real revolution had to be; how it had to go on from and to and fro between man and man, each one making his contribution. I was just another young man in a hurry. I thought I could carry the whole load, and I picked my dozen almost haphazard.

"What a crew they were! I am told that even these Gospels you talk about, are unflattering in their account of them.

"There is nothing flattering to be told about them. What a crew to start upon saving the world! From the first they began badgering me about their relative importance...

"And their *stupidity!* They would misunderstand the simplest metaphors. I would say, 'The Kingdom of Heaven is like so-and-so'... In the simplest terms...

"They always got it wrong.

"After a time I realised I could never open my mouth and think aloud without being misunderstood. I remember trying to make our breach with all orthodox and ceremonial limitations clear beyond any chance of relapse. I made up a parable about a Good Samaritan. Not *half* a bad story."

"We have the story," I said.

"I was sloughing off my patriotism at a great rate. I was realising the Kingdom of Heaven had to be a universal thing. Or nothing. Does your version go like that?"

"It goes like that."

"But it never altered their belief that they had come into the business on the ground floor."

105

"You told another good story about some Labourers in the Vineyard."

"From the same point of view?"

"From the same point of view."

"Did it alter their ideas in the least?"

"Nothing seemed to alter their ideas in the least."

"It was a dismal time when our great March on Jerusalem petered out. You know when they got us in the Garden of Gethsemane I went to pieces completely... The disciples, when they realised public opinion was against them, just dropped their weapons and dispersed. No guts in them. Simon Peter slashed off a man's ear and then threw away his sword and pretended not to know me...

"I wanted to kick myself. I derided myself. I saw all the mistakes I had made in my haste. I spoke in the bitterest irony. Nothing for it now but to know one had had good intentions. '*My* peace,' I said, 'I give unto you.'

"The actual crucifixion was a small matter in comparison. I was worn out and glad to be dying, so far as that went, long before those two other fellows – I forget who they were. One was drunk and abusive. But being crucified upon the irreparable things that one has done, realising that one has failed, that you have let yourself down and your poor silly disciples down and mankind down, that the God in you has deserted you – that was the ultimate torment. Even on the cross I remember shouting out something about it."

"Eli. Eli, lama sabachthani?" I said.

"Did someone get that down?" he replied.

"Don't you read the Gospels?"

"Good God, *No!*" he said. "How *can* I? I was crucified before all that."

"But you seem to know how things have gone?"

"It was plain enough how they were going."

"Don't you," I asked rather stupidly, forgetting where we were, "keep yourself informed about terrestrial affairs?"

"They crucify me daily," he said. "I know that. Yes."

2

Then without any sign of compunction, with that easy inconsistency which is so natural in the Dreamland atmosphere, he dropped the pose of knowing nothing of the Gospels and began to discuss them with the acutest penetration. He experimented with one explanation.

"People get here, good religious gentle beings, bringing the books they believe in. I talk to them because they are often so right-hearted that it perplexes me to find how wrong-headed they can be.

"Though I saw things going wrong after my crucifixion..."

That was not good enough. He went further and retold the story of the Resurrection...

"I saw that fellow Paul. That story is quite true. I fainted but I didn't die, and that dear old Joseph of Arimathea put me away in his own private sepulchre. Matthew's Gospel exaggerates about its being sealed and watched, and Mark, Luke and John came nearer the facts. I was put away by Joseph and Nicodemus, among a lot of spices and comforts, there was food and wine and fruit and even some money, and when I awoke I was disgusted

107

beyond measure to realise I was not dead. I sat there eating, because I was exhausted, and wondering what I had best do. Perhaps after all our Heavenly Father had a use for me, and, like yourself, I have never been willing to die. I would just obliterate myself for a time and think things over. You know something of that feeling."

"All my life," I said.

"Of course I ran up against some of the old set. There's no end of circumstantial truth in the gospels. Some of the women were hanging about the garden where I had been deposited. I didn't know what to say to them. I had no clear idea of the next step to take. I still felt there was much to be said and done in spite of the jumble I had made of it all. 'I must go away,' I said. 'Whither I go you cannot come. Stick to my teachings, and when I come back we'll have it all gloriously right.'

"Old Thomas thought I was a ghost, and he had to feel the nail holes in my hands before he would believe.

"I met two of them in an inn where I had gone for a bite, and we ate together. Gradually I got away from them as I worked my way north, to think things over from the beginning; while they got together to wait for that Second Coming.

"I was never much of a linguist – no Pentecost for me and so I couldn't go far into Syria beyond the range of my native Galilee."

"You didn't speak Hebrew?"

"Nobody *spoke* Hebrew in my time. Not a soul. It was a dead language. We used to read the Pentateuch and some of the prophets in the synagogues in Hebrew, and the scholars in Jerusalem kept up the cult, but the sort of Hebrew I knew was almost on a level with the rote-learnt Latin a provincial priest

108

would gabble in the early Dark Ages before the Benedictines bucked up classical learning. A majority of Jews, in Egypt for example, knew the Scriptures only through the Septuagint translation. When I got up and read the Law in the Nazareth synagogue, they realised my limitations and threw me out. This Aramaic we talked carried one far into Syria and along the coasts. So that I would wander away again when the Christians made trouble.

"I was not a bad carpenter, slow but careful and accurate. I rather liked Antioch – it was a big place in which one could lose oneself – if only Paul hadn't had a way of turning up there and asserting that I was the Son of the Holy Ghost. I knew there were some odd stories about mother. But you see I had every opportunity of hearing Paul on the top of his form. It was no good interrupting him. He would have bawled me out at once. And before I had thought out my problem, I died. I forget my last illness; some form of malaria, I think, and where my body is buried I haven't the faintest idea..."

"Not in the Holy Sepulchre?"

He smiled the Crusades away.

"And that's the true life and story of Jesus of Nazareth, the world's greatest failure?"

3

He began to change again. He smiled that charming confidential smile of his, which commonly preceded his leg-pulling phase, and then he became a different Jesus altogether, something much more Evangelical, the Pastor of a renascent City Temple perhaps. I did not know how to prevent that. You cannot dictate to a dream. He became – business-like. He escaped into the impossible.

"I don't know where you get your press cuttings," said this transfigured Jesus. "The stuff these infernal Christians pour out! I don't read a tenth of it."

Jesus of Nazareth reading press cuttings! "But *could* you *read*?" I protested, and woke up, before he could explain, as I know this metamorphosed Redeemer would have done, that he had been learning, learning.

But except for Dreamland the dead are finished and done. We have Holy Writ for it. My Jesus used to be fond of saying "It is written", but had that Jesus ever heard a single quotation from Ecclesiastes – with its stern insistence upon human finality?...

"A living dog is better than a dead lion... The dead know not anything... There is no work, nor device, nor knowledge, nor wisdom, in the grave, whither thou goest."

CHAPTER SIX

The Architect Plans the World

But that Carnival of the Gods and those bishops on broomsticks and that heart to heart talk about the difficulties of world-mending with one of its most celebrated failures, are just two among my endless adventures in Dreamland, and I do not know from night to night what new refreshment awaits me.

There is a lot of architecture beyond there. Sometimes I dream of a purely architectural world. But that architecture goes far beyond the mere putting-up of buildings and groups of buildings here and there. The architects of Dreamland lay out a whole new world. Their gigantic schemes tower to the stratosphere, plumb the depths of the earth, groom mountains, divert ocean currents and dry up seas. My identity merges inextricably with every dreamland architect. "We" do this; "we" do that. We share a common excitement at every fresh idea.

The agronomes – there are no farmers in Dreamland – come along and tell us, "We can produce all the food and the best of food and the most delightful food, not to mention all the drinks that make glad the heart of man, most easily and expeditiously for your happy thousands of millions, in those few hundred

111

thousand square miles we have marked out upon this globe for you, and the rest of the planet you can have to live in and make homes of and gardens of and playgrounds and – slightly controlled – wildernesses, and everything you architects can dream of and devise."

The geographers and metallurgists and mineralogists and engineers unfold their possibilities to us. "This is what you will be wise to do," they will say, "and this you can do if you will, and it is for you architects to see that none of our mines and pits become eyesores and offences against the ever acuter sensibilities of mankind," and forthwith we shall sit down with them and draw and redraw our plans. The artists will come demanding surfaces to decorate; the musicians will demand great soundproof auditoria, so that those who want to hear can hear and no one be bothered by unsolicited noise. All roads lead to architecture and building and rebuilding. These things We, the Creative King in Man, will carry out and carry on.

In these dreams I apprehend gigantic façades, vast stretches of magnificently schemed landscape, moving roads that will take you wherever you want to go instead of your taking them. "All this We do and more also," I rejoice. And though endless lovely new things are achieved, nothing a human heart has loved will ever be lost. I find myself a child again, the town-bred child I was, rejoicing in the sounds and sights of a country lane, delighting in by-ways where the honeysuckle twines about the ragged robin and one picks and nibbles the bread and cheese buds. Or one creeps through a hedge, conquering its resistance, into a tactfully unguarded garden where there are white raspberry canes and half-ripe gooseberries, black cherries and greengages. And then back to the grown-up magnificence once more...

Old fruits there are in Dreamland, but we feast on many marvellous new ones also. There is a vast Luther Burbank organisation at work upon them. And in these latter days, as the war effort strains our lives towards greater and greater austerity, there has been a notable increase of feasting and banqueting in my Dreams. We sit long at table, for there is time for everything where time ceases. I will not tantalise you with my last menu...

I cannot set any of these things down in sketches and forecasts and detailed descriptions, for how can I foretell what a hundred million brains, all better than mine, will conceive and plan and replan and continually enhance – they dissolve and vanish as I wake, but in my dream, I dream they are delightful beyond all experience, and, with that, Dreamland is satisfied.

CHAPTER SEVEN

Miracles, Devils and the Gadarene Swine

"These are the unavoidable distresses of a sick world. There is no way out of it. Every human being is a poisoned human being. We are all infected. A thousand contagions are in our blood. There is no health in us."

"And no magic remedy," we agree.

"Baptism? So natural and obvious to turn to that, and just symbolise our cleansing. So natural for poor infected things to hope and seek for Healers and Leaders to Well Being. So hard to tell them that there is no Salvation but in and through themselves."

"They used," said Jesus, detaching himself a little from me, "to crowd upon me – just as they would have crowded upon any passing novelty, gaping for me to do the tricks they called miracles. A lot of them were bed-rid malingerers and lazy spiritless people, but when I told them to get up and walk, with a threat in my eye, they walked all right. Also I made the disciples baptize and wash them. That proved an effective remedy in many cases. I never handled them myself. I've always had a sort of physical fastidiousness. Poor physique, I suppose...

"Beggars with good marketable sores used to get out of my way for fear of exposure. If sympathetic friends caught them and dragged them up to me, they had the alternatives of confessing themselves humbugs, which might have had disagreeable consequences for them, or adding to my thaumaturgic prestige. So they added to my thaumaturgic prestige.

"A lot of people in my time were possessed by devils, talked aloud, muttered queer things, frightened the timid and the children. Once they got that way with their neighbours, it was hard to get out of it. They liked being feared, of course, but they did not like being shunned and stoned. They appreciated the distinction of being possessed, but it is very inconvenient to be always possessed and never have a quiet moment. They found my stern exorcisms restful and acceptable. But many, when they realised that nobody was taking any notice of them, decided after a time to relapse and take unto themselves seven devils, each worse than the first…"

"The Gospels," I said, "are rather vague about that. They read rather as if you were just letting fly at the eagerness with which the enlightened relapse. Did you really believe there *were* devils?"

"Not finally. But at first yes. It was the prevalent idea and people lent themselves to it. There was that absurd affair with the pigs. Where was it?"

"The Gadarene swine? I'd love to hear about that."

"Yes, yes. One of my minor failures.

"The people came and begged me to stop my confounded miracles and get right out of the country as soon as possible. This fellow used to prowl about stark staring naked, exhibiting himself disgustingly, scaring girls and children assaulting people, howling and cursing and having a great time. At first everyone wanted me to do something about it. And *he* thought I could and would dispossess him. He did in a graveyard and when he realised I had got him, he came out in tremendous style. No solitary devil possessed him, he howled. Never was a man so bedevilled. He was a whole Legion. There were some pigs feeding near by and he made such an uproar that they took fright and stampeded down a steep place and into the sea. Quite a lot were drowned, and then it was the people came out and begged me to be off. They insisted. They saw me to my boat."

"I'm glad to have the story from you," I said. "As you know, the dear old Gospels are at sixes and sevens. Matthew, exaggerating as usual and spoiling his story, says there were *two* madmen, Mark says there was only one and that there were precisely two thousand pigs, while Luke tells a long story of how poor Legion wanted to come away with you out of the country. I can quite understand he felt he might be a bit unpopular... You left him behind."

"He was the aggressive sort of man *anyone* would leave behind," said Jesus. "Gladly."

"And you don't know what became of him?"

"I *think* they went up the hill to look for him after they had seen me off. They were business-like people, those Gadarenes. Whether they got him I don't know. The Gospels, you will have observed, tell nothing about it."

"Well, you had brought away all your Gospel witnesses."

2

"Those crowds! I used to insist upon the children coming nearest... They smelt – weaker. And they were artless. Yes – ... With or without a reason, I loved and pitied these foredoomed sacrifices to life. So long as they were little. It was a queer thing going from place to place, with something very urgent to say, that I knew ought to be said and which I was honestly trying to find words for, and to have to push my way through a smelly press of human degradation... All the time it was: 'Just one *little* miracle! Something we can tell our friends about. Something to *show!*'

"They would ask to have a wart or a whitlow cured – as a souvenir!

"A lot of them who had no luck with me couldn't bear the humiliation. So they went off and invented things. These downright liars just loaded me with wonders, far beyond anything else I did. They had a free hand..."

"That has been the common lot of everyone who felt there was something he ought to say," I remarked. "Everyone. Buddha and St Benedict, every saint in the calendar is half buried under a cairn of marvel-mongers' lies. Mankind would have smothered itself in its own lies long ago, if history were not so plainly incredible. Truth has a way of heaving up through the cracks of history. Or we should be damned without hope."

3

We sat digging our toes into the Elysian greensward, reflecting, in the infinite leisure of finished lives, upon our particular failures to release the human thought that still seeks release and realisation in time and space.

"Your miracle cadgers remind me of autograph hunters," I remarked.

"Never heard of them," he said. "What are they?"

"They leave you alone at ordinary times, and they haven't the slightest desire to hear a word you have to say at any time, but if you to make a lecture or a broadcast, they seem to spring up like flies from offal and buzz about your face. But when you try to tell them the truth of things, they say you are 'preaching' and turn away."

He sat judging his past impartially.

"It was hard not to humbug just a bit to catch their attention… Just *manage* them a bit. Once when I had been feasting in the desert and trying to get my ideas in order, I saw the Devil in a sort of vision."

"Jehovah's Satan?" I asked, delighted. "The actual Tempter?"

"The very Devil. That was how the vision arranged itself. He *was* a brisk plausible conceited ass, a lot subtler than poor old 'Thou-Shalt-have-none-other-Gods-but-Me'. He took me up some sort of pinnacle. 'If only you would let me manage things for you,' said he, 'you could do anything you liked with the world – anything. All you have to do is to put yourself in my hands as your impresario – and just *say* anything you please.' "

"You said: 'Get thee behind me, Satan.' "

"My expression was more colloquial.

" 'Have your own way,' he said, and vanished, and down I came to find the multitudes, mostly with furtive lunch packets, coming out to look for me. They had no confidence in my power to exalt them above hunger and thirst. They expected miracles, but not that sort, and when they found there was food to burn, they were ashamed of themselves and pretended not to have brought anything. So that again was put down to my miraculous gifts...

"Hard not to take Satan's suggestion and avail oneself of his artful dodges... Just what one must not do. Mankind must learn together and rise together. Mankind is one. *We* are one."

"Mankind is learning," we said. "We have schools here..."

We thought in perfect unison. We had suddenly coalesced.

"Let us go and see them," said we. "They won't notice us. We are just wraiths from the frustrated past. They will walk through us and never know we are there."

CHAPTER EIGHT

A Hymn of Hate Against Sycamores

Only so many hours a day can one work full out, and only for so many may one eat, digest or sleep, and the strange forces that make us what we are, as the price of an efficient discharge of these functions, insist upon certain performances called exercises, whereby we are compelled to scramble up and down steep places, row boats, box, fence, dance, ride oscillating horses, hit balls about, dig in gardens or otherwise provoke our perspiration pores to a copious activity – involving immediate change of raiment or rheumatism.

In my efforts to maintain and protect a satisfactory thread of thought and purpose for myself and my species amidst this labyrinth of ridiculous animal servitudes, I have achieved considerable economies of time and energy. One main meal a day is better than several, and all that walking and running about with rackets and golf clubs and ball-hitting implements can be eliminated, I find, by concentrating upon one's garden. Man, says Holy Writ, was created a gardener. He yields to the suggestion very readily.

The honest tradesman retires from business with gardening in his mind; he may take a side interest in bowls or suchlike elderly sports, but it is the officially mild obedient yielding garden that dominates his released imagination. He is still much preoccupied by the endless mischievous waywardness of this new sphere he has acquired, when he in his turn is dibbled out, like so many of his seedlings, in the hope rather that the certainty of a glorious resurrection...

Candide experiences all the variety of life, and, disillusioned with everything else, comes back to the service of these smug impostures upon our planet...

I simply follow the disposition of my kind, when, between the comprehensive efforts to restate existence in terms of space-time, that engage my dwindling hours of maximum validity, I turn from the Macrocosm to the microcosm and carry on, with an intensifying interest, a struggle, that under the stresses of our present limitations becomes more and more single-handed and inexpert, against the indisciplines, contradictions and disorders of my awkward squad of shrubs and plants and flowers. It is such a *little* back garden; a garden to be laughed at, a small paved garden with beds on either side which used to be filled with flowering plants by a nurseryman, who replaced them by others when they were over. It never dawned upon me that it was also a Cosmos, until I was left alone with it. But now, trowel in hand, I know there is no real difference except in relative size between my seedling antirrhinums and the Pleiades. They both mock me – in a parallel manner.

And among the many things I have learnt from this microcosm is the incredible fierceness, nastiness and brutality of the Vegetable Kingdom. It feeds greedily upon filth in any form, and its life is wholly given up to torture and murder. It is a common delusion that there is something mild and moral about all this green stuff in which our planet is clothed. It is really a question of the pace at which one sees it. We, the higher animals, scurry through life so fast that we do not note the more deliberate horrors of these plant lives. It is only now and then, in a jungle, or amidst the towering white menace of a burnt or burning Australian forest, that Nature strips the moral veils from vegetation and we apprehend its stark ferocity. I doubt if Voltaire ever came face to face with that garden of his without some intervening help; he wrote the end of *Candide* and died before he realised the truth.

In this back garden of mine, this little Creation to which I play the Lord, I see seedling, bush and tree attacking each other pitilessly and relentlessly, and in particular I have watched a very delightful little almond-tree I loved, done to death before I could do anything to save it. Nearest of the murderers was a holly-tree which I have now sawn almost to the ground and would have destroyed altogether except that a curious albino sport springs from its root; certain rivals for my affection contributed to the outrage; but the chief of the slaughter gang, and still an ever increasing affliction of all I would keep gentle, healthy and beautiful, is a vast lumping Sycamore that grows in the deserted garden next door to me. Like most of my erstwhile neighbours in this Hanover Terrace, next-door has gone away, but he retains his lease; his abandoned garden is a centre of weed distribution, and before I can get anything done about it, all sorts of authorities have to be consulted, and after that I doubt if it will be possible to find the labour necessary to terminate the ever spreading aggressions of this hoggish arborial monster.

Every day when I go out to look at my garden I shake my fist at it and wish for the gift of the evil eye. Every day it grows visibly larger, ignoring my hatred. It is not only my garden it devastates. It is destroying the gardens beyond, which also are now abandoned. There there are laburnums and acacias and many abandoned helpless flowering shrubs and plants, awaiting their doom.

These Sycamores are not even entitled to be called Sycamores; they have assumed the name of a better plant than themselves. Says my *Oxford English Dictionary;* "It is commonly spoken of with a distinguishing adjective as the 'bastard, base or vulgar' sycamore." The true sycamore, or sycomore, as the Bible has it, was a mulberry- or fig-tree (sycon = Fig), and the people who bear the name in English are probably immigrants from more classical parts of the world who would have done better to call themselves Mulberries instead of mixing themselves up with this vile blob upon the English landscape. In all that follows, when I write "Sycamore", I do not point at them. This tree of mud has stolen their name, and I invite them to share my hate and indignation.

I am trying to find out what scoundrel first brought these gawky trees to my England and my London. It was the work of a cheapener, a fundamentally dishonest spirit. They grow with great rapidity; they can stand the now diminishing London soot because they have a stolid will to live through anything, no matter what evil ensure. So that to the cheap and nasty building estate practitioner they were easy to pass off as even desirable trees. They were the perfect tree for the suburban jerry-builder. They seed pitilessly, and they disseminate minute irritant hairs very bad for the respiratory passages. In leaf they are as blowsy as tippling charwomen, and even when they are stripped they have as little allure. They are more like the compositions of

Vaughan Williams or Eric Coates than anything else I know. They branch out with a stupid elimination of the unexpected. You never say to a sycamore as you do to all good music and all lovable trees: "Of course that sequence is exactly right, but who could have thought of it, before you did it!"

A Sycamore, if you told it that, would be disconcerted and wonder whether it had not made some sort of slip...

Men marry Sycamores and by all our laws they are blameless women. It is no plea for a divorce in this preposterous world of ours that a wife has an infinite want of variety.

Well, there is this dirty, ugly, witless, self-protective tree, blighting London; it is everywhere, and I hear no voice raised against it. The other day I went to see an exhibition of designs for the rebuilding of London, and there I saw Sycamores as men walking, and they were scheming as awful a London of squalid jobbery as it is possible to imagine. They just wanted to grow all over it abundantly, sycamore-fashion. They were too stupid to have an idea of the New World we poor sensitive men dream of extracting from our present disasters. They did not know and they did not care whether the world was to be a world at peace or a world at war, living underground in perpetual fear of blitzkriegs or towering up to the skies, a world of universal free trade or shabby competition. They did not know and they did not care. They had schemes for green belts – mostly of sycamore-trees – and rows and rows of nasty little building estate houses. They were out for jobs, they were unable to imagine any other jobs except squalid enlargements of their own, and the blight of their dullness fell upon me, so that I came home cursing and swearing, to the dismay of passers-by, and now, whenever I look over my garden wall at this vast dreary tree, waving its leaves at me, I see them also. Every morning that tree seems to come nearer to me and overhang me more and more.

The Sycamore is a complete repudiation of any belief in an intelligent God. One may perhaps believe in a God who made good and evil, but the creation of these Sycamores, men, women and trees, was just damned stupidity.

I shall fight evil to my last breath because that is my nature, but it is the thought of these Sycamores that brings me nearest to despair.

So let me conclude by cursing Sycamores and all who favour and abet Sycamores and have sycamore elements in their nature, and let me avoid all vulgar and irreligious cursing, and curse strictly in the terms provided in Holy Writ, in the Twenty-eighth Chapter of the Book of Deuteronomy.

Listen all ye of the Sycamore tribe, and thank your lucky stars, Mr Mulberry Sycamore, that it does not apply to you!

I draw a deep breath and indicate the same by a white line.

"Cursed shalt thou be in the city, and cursed shalt thou be in the field.

"Cursed shall be thy basket and thy store.

"Cursed shall be the fruit of thy body, and the fruit of thy land, the increase of thy kine, and the flocks of thy sheep.

"Cursed shalt thou be when thou comest in, and cursed shalt thou be when thou goest out.

"The Lord shall send upon thee cursing, vexation and rebuke, in all that thou settest thine hand unto for to do, until thou be destroyed, and until thou perish quickly; because of the wickedness of thy doings...

"The Lord shall make the pestilence cleave unto thee, until he have consumed thee from off the land, whither thou goest to possess it.

"The Lord shall smite thee with a consumption, and with a fever, and with an inflammation, and with an extreme burning, and with the sword, and with blasting, and with mildew; and they shall pursue thee until thou perish.

"And thy heaven that is over thy head shall be brass, and the earth that is under thee shall be iron.

"The Lord shall make the rain of thy land powder and dust: from heaven shall it come down upon thee, until thou be destroyed...

"And thy carcase shall be meat unto all fowls of the air, and unto the beasts of the earth, and no man shall fray them away.

"The Lord will smite thee with the botch of Egypt, and with the emerods, and with the scab, and with the itch, whereof thou canst not be healed.

"The Lord shall smite thee with madness, and blindness, and astonishment of heart:

"And thou shalt grope at noonday, as the blind gropeth in darkness, and thou shalt not prosper in thy ways: and thou shalt be only oppressed and spoiled evermore, and no man shall save thee.

"Thou shalt betroth a wife, and another man shall lie with her: thou shalt build a house, and thou shalt not dwell therein: thou shalt plant a vineyard, and shalt not gather the grapes thereof.

"Thine ox shall be slain before thine eyes, and thou shalt not eat thereof: thine ass shall be violently taken away from thee before thy face, and shall not be restored to thee: thy sheep shall be given unto thine enemies, and thou shalt have none to rescue them.

"Thy sons and thy daughters shall be given unto another people, and thine eyes shall look, and fail with longing for them all day long: and there shall be no might in thine hand...

"The Lord shall smite thee in the knees, and in the legs, with a sore botch that cannot be healed, from the sole of thy foot unto the top of thine head...

"Thou shalt carry much seed out into the field, and shalt gather but little in; for the locust shall consume it.

"Thou shalt plant vineyards, and dress them but shalt neither drink of the wine, nor gather the grapes; for the worms shall eat them.

"Thou shalt have olive trees throughout all thy coasts, but thou shalt not anoint thyself with the oil; for thine olive shall cast his fruit.

"Thou shalt beget sons and daughters, but thou shalt not enjoy them; for they shall go into captivity.

"All thy trees and fruit of thy land shall the locust consume.

"The stranger that is within thee shall get up above thee very high; and thou shalt come down very low.

"He shall lend to thee, and thou shalt not lend to him; he shall be the head, and thou shalt be the tail.

"Moreover all these curses shall come upon thee, and shall pursue thee, and overtake thee, till thou be destroyed.

"Because thou servedst not the Lord thy God with joyfulness, and with gladness of heart, for the abundance of all things;

"Therefore shalt thou serve thine enemies which the Lord shall send against thee, in hunger, and in thirst, and in nakedness, and in want of all things: and he shall put a yoke of iron upon thy neck, until he shall have destroyed thee.

"And thou shalt eat the fruit of thine own body, the flesh of thy sons and of thy daughters, which the Lord thy God hath given thee, in the siege, and in the straitness, wherewith thine enemies shall distress thee:

"So that the man that is tender among you, and very delicate, his eye shall be evil toward his brother, and toward the wife of his bosom, and towards the remnant of his children which he shall leave:

"So that he will not give to any of them of the flesh of his children whom he shall eat: because he hath nothing left him in the siege, and in the straitness, wherewith thine enemies shall distress thee in all thy gates...

"Then the Lord will make thy plagues wonderful, and the plagues of thy seed, even great plagues, and of long continuance, and sore sicknesses, and of long continuance.

"Moreover he will bring upon thee all the diseases of Egypt, which thou wast afraid of; and they shall cleave unto thee.

"Also every sickness, and every plague, which is not written in the book of this law, them will the Lord bring upon thee, until thou be destroyed."

That last clause is a magnificent piece of curse drafting. Not a loophole remains.

This, I think, is all that needs to be said about my neighbour's Sycamore in particular and Sycamores and Sycamorism in general. I can imagine nothing more comprehensive. I can add nothing to it. Take it, Mr Sycamore, take it all and be damned to you! And thank the powers of earth and heaven, Mr Mulberry Sycamore, that it is not to you that these words are addressed.

This cursing, let us realise, is the sort of thing the Pope, his Cardinals, the Archbishops, Bishops, priests and deacons, the pious controllers of the BBC, and all the Sycamore Groves of canting Christendom, declare they find so good for the soul of man. This is the spirit of the Sacred Book they distribute about the world to teach men love and gentleness.

The plain, if inadvertent, evidence of Holy Writ is that from the beginning, God knew he had made a mess of things and set Himself to savage his Creation. Time after time, he repented that he had made man, and time after time he sent floods and judgments. He seems to have found Creation almost as obdurate and frustrating and exasperating as I do in my garden.

Here in the freedom of Dreamland I recognise and deal with these Christian teachers for the foolish weaklings they are. I refuse to accept this consecrated riff-raff as my moral and mental equals. Clearly they are either knaves or fools or a blend in various proportions of the two, and to treat them as though they were intelligent honest men in this world crisis, becomes a politeness treasonable to mankind.

I find this little outburst a great relief.

(Thank you, God! For if you serve no other purpose in Dreamland you are still admirable to swear by.)

CHAPTER NINE

The Divine Timelessness of Beautiful Things

T he other day the Happy Turning took me to the sunlit sweetness of the Elysian fields, and sometimes, after the manner of Dreamland, it seemed to me I was talking to a great number of poets, painters, artists, makers of every sort, and sometimes that I was just talking to myself, and the talk was all about the beautiful things that man has got out of this unrighteous world, and whether there can ever be another happy harvest of Beauty, and if so, what sort of harvest it may be.

A point we found we were all agreed upon was that Beauty is eternal and final, a joy for ever. There is no progress in it and no decline. You cannot go beyond it. You may make replicas of it; you may record and imitate it, you can destroy it for yourself and others, obliterate it and blaspheme it, but you cannot do away with its invincible divinity. Even when it is a lost God, a Beauty is still God, a being in itself, serene, untroubled, above all the accidents of space and time.

But what we had most in mind was this, that there is a definite limit set to the abundance of any particular Beauty. It is discovered, it is revealed, and that is its end. That God has smiled and passed and returns no more. Other Gods may smile in their turn, and they too will pass away.

We cited instances of these immortal visitations.

There was, said a classical scholar, that gracious beauty which was distilled by the Hellenic poets and sculptors out of the vast confusion of antique mythology. It has lit this dull world for all its lovers with an inalienable charm. Pan and the dryads haunt the woodlands, the naiads bathe in the stream, Diana steals down the beams of misty silver to Endymion, and eternally amidst the glittering waters, Triton blows his wreathèd horn.

"But one thing goes on," said a man who called himself an anthologist, "and that is the creative magic in English poetic creation." Which threw us all into an intricate disputation that carried us over the whole field of English literature and drama and was shot with a flashing multitude of interests and surprises. "There is not one single Goddess here," we agreed, "but a varied sisterhood, and most of these sisters are wantons and have led lives that make the Olympians seem by comparison calm and consistent and at least superficially decorous." Gradually we begin to disentangle the preoccupations of these lively Beauties.

There is that lost Goddess of beautiful English who, with little Latin and less Greek, played with it so delightfully in Shakespearian days and was finally murdered by her Latin lover in a fit of jealousy because she flirted with the far more lively colloquial scullion downstairs. She came to her tragic end before the Stuarts were done for. For a while she lay calm and rigid in death before her ultimate decay. All that Swift and Sterne, Addison and Gray and Gay, albeit they loved her greatly, could achieve was an unexciting pellucid flow. The *Dunciad* is the dirge of a happy lovely language lying dead under a black pall of Hanoverian gutturals.

Dear heart! she left one bastard by philosophy, not a Goddess indeed but a demi-Goddess, the Wordsworthian discovery of the mystical loveliness beneath reality, but for the rest, we Dreamland anthologists asked, what later Beauty of English is worth our keeping? Newdigate prizewinners, pompous and pretentious verse-makers, the massive unpired industrious professionalism of Tennyson, head expert of the industry, Longfellow doing his level best, and never succeeding, to make Laughing-Water Hiawatha laugh, the fumed oak stuff from the Morris antique shop, the vanity, crudity and unimaginative topicality of that overrated etcher, Blake, the jingling vulgarities of Byron, Martin Tupper, Alfred Noyes, T S Eliot, Bridges and the rest of them – as void of the mysterious exaltation of Beauty as a crew of disinherited mourners at a bankrupt's funeral on a wet day. Who in the great world we dream about will delight in any of this later stuff? Have we any use for it at all?

The anthologist did his best. "There are *bits*," he pleaded, digging nervously in the addled egg, that curate's egg, of later English poetry. "A rose-red city half as old as time," he quoted, but he could not recall the name of the man who produced that one happy line, and then he bethought himself suddenly of Shelley.

He dredged up a few quotable lines. "The earth doth like a snake renew its winter skin outworn." And a fragment of *Queen Mab*.

"Well?" he said.

"You shall have that," I conceded, "though much of Shelley is copious, intellectualised and tedious stuff, last bubbles from the

133

drifting body of the drowned Goddess, and, such as they were, they rose to the surface and broke and vanished a century and a quarter ago. But all the rest was just trying to go on with something that indeed was finished for ever."

It was my Dream, entirely mine for a while; no one said anything more; and thus, having left English poesy for dead, these fluctuating dream Elysians fell to discussing one of the most radiant smiles of another of these – wanton English Beauties – who lived so fast and gaily in those days of literary loveliness, the divine imagination of the *Midsummer Night's Dream*.

There we agreed, was a piece of the magic that can never lose its charm. "Or *The Tempest*," said someone.

There again we had a culminating finish, something done, for good and all, so that nothing of the same supreme sort could ever be done again. But all through that happy phase of English inspiration beauty flashed and quivered. Brightest among the London Globe galaxy who slapped together the plays and poetry which people now call "Shakespeare", was a brilliant youth of that name, who loved all too freely and retired to Stratford on Avon to die untimely, as his final signatures show, in the mute misery of incipient GPI. He stood out among them all in his early years, but, quite apart from these distinctive creations, the language was in such a state that hardly anyone could touch it without striking sparks of loveliness, and one must be very heavily erudite to attribute any particular single flash in the collection to this man or that.

And as My Dreamland company talked in Elysium we became aware of a curious unanimity about that respectable triology, the Beautiful, the Good and the True. Dear old Professor Gilbert

Murray appeared among the eternal sunlit greenery and was greeted with a respectful murmur. He declared with a defiant flash in his glasses and a note of passion in his voice, that he believed in the Good, the Beautiful and the True, but, before he could be asked any questions, he vanished from among us completely, and we were left to consider what he meant by these words. We found we were agreed that he had put three realities, essentially different in their nature, upon a quasi-equality for which there was no justification whatever.

We left the moral factor, the Good, aside for the present. Goodness is a matter of *mores*, of good social behaviour, and there is so wide a diversity of social values in the world that it seemed unnecessary to my Elysians to question anything so impermanent. The transitoriness of morality is in flat contrast to the deathless finality of beauty.

But when we turned to literature which does not pretend to beauty in the first place, but to interest of statement or narrative, we found something, that only verges, as it were, in a few incidental passages, and by accident, on poetic beauty. For the rest, literature, both the philosophical, the "scientific" and the fictitious, is telling about what things are, what life is; about its excitements, its emotional effects, its expectations, its laughter and tears. It is as different from poesy as apple pie is from Aurora.

This work of the human mind in telling and enforcing a view has produced a huge real literature quite apart from the infinitely vaster sham literature which is foisted upon people whose cacotrophically educated undiscriminating minds cannot ever perceive they are being told nothing at all, and who read in a muzzy fashion, as people play patience, because they can think of nothing better to do.

This literature of reality has not the permanence of beauty. It absorbs and reproduces the storytelling and statements of the past. It does its utmost to recapture from the past the experiences swallowed by the maw of time. Or it invents typical or experimental characters to try over problems and variations in conduct. There can be no classical novels or romances. The strictly circumstantial ones last longest. Fielding's *Voyage to Lisbon* will outlive *Tom Jones*. Stories become tedious as our vision broadens. Nor are there classics of science. Knowledge pours in continually to amplify and correct. Yet every new realisation, every fresh discovery, has for those who make it, a quality of beauty, transitory indeed but otherwise as clear and pure as that enduring Beauty we cherish for ever, and ephemeral beauty for one man or for a group of mortals, sufficient to make a life's devotion to the service of truth worth while.

So we found ourselves in agreement that the human mind may be in a phase of transition to a new, fearless, clear-headed way of living in which understanding will be the supreme interest in life, and beauty a mere smile of approval. So it is at any rate in the Dreamland to which my particular Happy Turning takes me. There shines a world "beyond good and evil", and there, in a universe completely conscious of itself, Being achieves its end.

H G Wells

The History of Mr Polly

Mr Polly is one of literature's most enduring and universal creations. An ordinary man, trapped in an ordinary life, Mr Polly makes a series of ill-advised choices that bring him to the very brink of financial ruin. Determined not to become the latest victim of the economic retrenchment of the Edwardian age, he rebels in magnificent style and takes control of his life once and for all.

ISBN 0-7551-0404-8

H G WELLS

IN THE DAYS OF THE COMET

Revenge was all Leadford could think of as he set out to find the unfaithful Nettie and her adulterous lover. But this was all to change when a new comet entered the earth's orbit and totally reversed the natural order of things. The Great Change had occurred and any previous emotions, thoughts, ambitions, hopes and fears had all been removed. Free love, pacifism and equality were now the name of the game. But how would Leadford fare in this most utopian of societies?

ISBN 0-7551-0406-4

H G WELLS

THE INVISIBLE MAN

On a cold wintry day in the depths of February a stranger appeared in The Coach and Horses requesting a room. So strange was this man's appearance, dressed from head to foot with layer upon layer of clothing, bandages and the most enormous glasses, that the owner, Mrs Hall, quite wondered what accident could have befallen him. She didn't know then that he was invisible – but the rumours soon began to spread...

H G Wells' masterpiece *The Invisible Man* is a classic science-fiction thriller showing the perils of scientific advancement.

ISBN 0-7551-0407-2

H G WELLS

THE ISLAND OF DR MOREAU

A shipwreck in the South Seas brings a doctor to an island paradise. Far from seeing this as the end of his life, Dr Moreau seizes the opportunity to play God and infiltrate a reign of terror in this new kingdom. Endless cruel and perverse experiments ensue and see a series of new creations – the 'Beast People' – all of which must bow before the deified doctor.

Originally a Swiftian satire on the dangers of authority and submission, Wells' *The Island of Dr Moreau* can now just as well be read as a prophetic tale of genetic modification and mutability.

ISBN 0-7551-0408-0

H G Wells

Men Like Gods

Mr Barnstaple was ever such a careful driver, careful to indicate before every manoeuvre and very much in favour of slowing down at the slightest hint of difficulty. So however could he have got the car into a skid on a bend on the Maidenhead road?

When he recovered himself he was more than a little relieved to see the two cars that he had been following still merrily motoring along in front of him. It seemed that all was well – except that the scenery had changed, rather a lot. It was then that the awful truth dawned: Mr Barnstaple had been hurled into another world altogether.

How would he ever survive in this supposed Utopia, and more importantly, how would he ever get back?

ISBN 0-7551-0413-7

H G Wells

The War of the Worlds

'No one would have believed in the last years of the nineteenth century that this world was being watched keenly and closely by intelligences greater than man's…'

A series of strange atmospheric disturbances on the planet Mars may raise concern on Earth but it does little to prepare the inhabitants for imminent invasion. At first the odd-looking Martians seem to pose no threat for the intellectual powers of Victorian London, but it seems man's superior confidence is disastrously misplaced. For the Martians are heading towards victory with terrifying velocity.

The War of the Worlds is an expertly crafted invasion story that can be read as a frenzied satire on the dangers of imperialism and occupation.

ISBN 0-7551-0426-9

OTHER TITLES BY H G WELLS AVAILABLE DIRECT
FROM HOUSE OF STRATUS

Quantity	£	$(US)	$(CAN)	€
FICTION				
ANN VERONICA	9.99	14.95	22.95	16.50
APROPOS OF DOLORES	9.99	14.95	22.95	16.50
THE AUTOCRACY OF MR PARHAM	9.99	14.95	22.95	16.50
BABES IN THE DARKLING WOOD	9.99	14.95	22.95	16.50
BEALBY	9.99	14.95	22.95	16.50
THE BROTHERS AND				
THE CROQUET PLAYER	7.99	12.95	19.95	14.50
BRYNHILD	9.99	14.95	22.95	16.50
THE BULPINGTON OF BLUP	9.99	14.95	22.95	16.50
THE DREAM	9.99	14.95	22.95	16.50
THE FIRST MEN IN THE MOON	9.99	14.95	22.95	16.50
THE FOOD OF THE GODS	9.99	14.95	22.95	16.50
THE HISTORY OF MR POLLY	9.99	14.95	22.95	16.50
THE HOLY TERROR	9.99	14.95	22.95	16.50
IN THE DAYS OF THE COMET	9.99	14.95	22.95	16.50
THE INVISIBLE MAN	7.99	12.95	19.95	14.50
THE ISLAND OF DR MOREAU	7.99	12.95	19.95	14.50
KIPPS: THE STORY OF A SIMPLE SOUL	9.99	14.95	22.95	16.50
LOVE AND MR LEWISHAM	9.99	14.95	22.95	16.50
MARRIAGE	9.99	14.95	22.95	16.50
MEANWHILE	9.99	14.95	22.95	16.50
MEN LIKE GODS	9.99	14.95	22.95	16.50
A MODERN UTOPIA	9.99	14.95	22.95	16.50
MR BRITLING SEES IT THROUGH	9.99	14.95	22.95	16.50

ALL HOUSE OF STRATUS BOOKS ARE AVAILABLE FROM GOOD BOOKSHOPS
OR DIRECT FROM THE PUBLISHER:

Internet: **www.houseofstratus.com** including synopses and features.

Email: **sales@houseofstratus.com**
info@houseofstratus.com
(please quote author, title and credit card details.)

OTHER TITLES BY H G WELLS AVAILABLE DIRECT
FROM HOUSE OF STRATUS

Quantity	£	$(US)	$(CAN)	€
FICTION				
THE NEW MACHIAVELLI	9.99	14.95	22.95	16.50
THE PASSIONATE FRIENDS	9.99	14.95	22.95	16.50
THE SEA LADY	7.99	12.95	19.95	14.50
THE SHAPE OF THINGS TO COME	9.99	14.95	22.95	16.50
THE TIME MACHINE	7.99	12.95	19.95	14.50
TONO-BUNGAY	9.99	14.95	22.95	16.50
THE UNDYING FIRE	7.99	12.95	19.95	14.50
THE WAR IN THE AIR	9.99	14.95	22.95	16.50
THE WAR OF THE WORLDS	7.99	12.95	19.95	14.50
THE WHEELS OF CHANCE	7.99	12.95	19.95	14.50
WHEN THE SLEEPER WAKES	9.99	14.95	22.95	16.50
THE WIFE OF SIR ISAAC HARMAN	9.99	14.95	22.95	16.50
THE WONDERFUL VISIT	7.99	12.95	19.95	14.50
THE WORLD OF WILLIAM CLISSOLD				
VOLUMES 1,2,3	12.99	19.95	29.95	22.00
NON-FICTION				
EXPERIMENT IN AUTOBIOGRAPHY				
VOLUMES 1,2	12.99	19.95	29.95	22.00
H G WELLS IN LOVE	9.99	14.95	22.95	16.50
THE OPEN CONSPIRACY AND				
OTHER WRITINGS	9.99	14.95	22.95	16.50

Tel:	Order Line 0800 169 1780 (UK) 1 800 724 1100 (USA)	International +44 (0) 1845 527700 (UK) +01 845 463 1100 (USA)
Fax:	+44 (0) 1845 527711 (UK) +01 845 463 0018 (USA) (please quote author, title and credit card details.)	
Send to:	House of Stratus Sales Department Thirsk Industrial Park York Road, Thirsk North Yorkshire, YO7 3BX UK	House of Stratus Inc. 2 Neptune Road Poughkeepsie NY 12601 USA

PAYMENT (Please tick currency you wish to use):

☐ £ (Sterling) ☐ $ (US) ☐ $ (CAN) ☐ € (Euros)

Allow for shipping costs charged per order plus an amount per book as set out in the tables below:

CURRENCY/DESTINATION

	£(Sterling)	$(US)	$(CAN)	€(Euros)
Cost per order				
UK	1.50	2.25	3.50	2.50
Europe	3.00	4.50	6.75	5.00
North America	3.00	3.50	5.25	5.00
Rest of World	3.00	4.50	6.75	5.00
Additional cost per book				
UK	0.50	0.75	1.15	0.85
Europe	1.00	1.50	2.25	1.70
North America	1.00	1.00	1.50	1.70
Rest of World	1.50	2.25	3.50	3.00

PLEASE SEND CHEQUE OR INTERNATIONAL MONEY ORDER
payable to: HOUSE OF STRATUS LTD or HOUSE OF STRATUS INC. or card payment as indicated

STERLING EXAMPLE

Cost of book(s):. . . . . Example: 3 x books at £6.99 each: £20.97

Cost of order: . . . . Example: £1.50 (Delivery to UK address)

Additional cost per book:. . . . . Example: 3 x £0.50: £1.50

Order total including shipping:. . . . . Example: £23.97

VISA, MASTERCARD, SWITCH, AMEX:

☐☐☐☐☐☐☐☐☐☐☐☐☐☐☐☐☐☐

Issue number
(Switch only):   **Start Date:**   **Expiry Date:**

☐☐☐   ☐☐/☐☐   ☐☐/☐☐

Signature: _____

NAME: _____

ADDRESS: _____

COUNTRY: _____

ZIP/POSTCODE: _____

Please allow 28 days for delivery. Despatch normally within 48 hours.
Prices subject to change without notice.
Please tick box if you do not wish to receive any additional information. ☐

House of Stratus publishes many other titles in this genre; please check our
website (**www.houseofstratus.com**) for more details.